A Qualitative & Quantitative Developmental Math Assessment
and
Intervention Protocol

By Lisa Ann de Garcia

A Qualitative and Quantitative Developmental Math Assessment and
Intervention Protocol.

A portion of the materials contained in this publication was created with the use
of **1, 2, 3 Math Fonts**

Table of Contents

Fractions

+ Acknowledgement

I have been greatly influenced by the many pioneers in mathematics education, including, but not limited to, Kathy Richardson, Marilyn Burns, Thomas Carpenter, Kathy Fosnot, Ruth Parker, Eula Monroe, and Damon Bahr. Learning how to ask questions and listen to my students' thinking transformed my teaching. My ability to work with the struggling learner was enhanced once I understood the developmental phases that students pass through as they are learning about number and operations, stemming from the works of *First Steps* from Australia and Kathy Richardson. When I was in the classroom, I knew I had students who were struggling, but I did not know how to pinpoint their weaknesses or what to do about them. Now I am able to trace their difficulties as either rooting in a lack of early numeracy or place value understanding. I would like to acknowledge that this assessment was inspired from Kathy Richardson's *Assessing Math Concepts, Steps* Professional Development from Australia, the *Math Reasoning Inventory* by Marilyn Burns, and the Extended Day Mathematics Intervention Program developed by Brian Tash and others at San Diego City Schools. If you would like to learn how to provide more in depth assessments in specific developmental areas, please refer to the resources by Kathy Richardson at www.didax.com, for primary, and the free resources developed by Marilyn Burns www.mathreasoninginventory.com for upper grade students.

I have been equally in awe of neuroscience and the application it has on education. The current understanding in brain research that the brain is plastic and can learn given the right stimulus, has been revolutionary. I continue to research how specific learning activities can impact the brain. I am grateful to the children whose lives I have been fortunate to touch, as they have taught me more than I could ever teach them. I have been very blessed to have been able to see children's mathematical lives transformed. It is all worth it when children have their first "ah ha" moment, smiles for the first time in math, and now says it is their favorite time of the day.

I cannot fail to acknowledge my own family. Each of my 3 boys has taught me very important lessons about how children learn. One, a very right brain learner who has needed a lot of support in our left brained educational system, another who has autism and I have had to teach him math as if he were a second language learner, and a third who appears to have turned out to be a little more like me...a lover of knowledge. Each one has his own brilliance, and all three have had to put up with me experimenting on them and videotaping their thinking. Finally, my husband who has put up with me all of these years just being me.

Overview

This assessment is a result of working with small groups and individual students who were struggling in mathematics. I needed a way to quickly identify the developmental level of my students in the area of number and operation. Using classroom test data was not enough, since struggling students can be at least 2 years below grade level. I needed to see exactly where they were so I could help them move on to the next developmental phase, not just try to cope with what was happening in the classroom. I also needed a way to track progress since growth may not transfer to standardized tests for a couple of years, even though they may be making developmental improvements. I wanted a single assessment that I could easily use with any child ranging from first through sixth grades.

I noticed that I tended to use similar lessons or activities with children who fell within specific developmental ranges. I decided to document these key activities hoping that they could be a resource to other teachers working with struggling children. The length of time I spend on an activity, or series of activities, vary depending on the individual students. Many struggle with retention and need a lot more exposure to concepts than other students. I am also coming up frequently with new and creative ways to address a concept, so obviously teachers bring with them their own bag of tricks and creativity. It is impossible to include everything that could be done to support one of the concepts tested in this assessment, but I included the ones I felt were most significant.

The Assessment

1-1 student assessments are quite qualitative in nature, however, in today's environment where the use of data is highly valued, I wanted a way to quantify what I was noticing. Many skills are scored on a 0-3 rubric, and others are simply worth 1 point. I found that I was able to easily track the learning of my students this way and was able to justify the methods I was using by documenting growth in a tangible way.

The assessment starts with early number concepts and goes through computation with fractions. I find that the majority of my students fell somewhere right around the understanding of place value. The test is naturally divided into three segments. The first segment I use for all of my students, then I determine whether to give the primary or upper version. I generally give the primary version if my student is in second grade or younger. Unless I know something unique about my student, I will generally start third graders with the upper version. If they are unsuccessful with the beginning, I will go back to the primary version.

Assessment overview:

All students: Combination of number to ten
10 and some more
Reading and Comparing Numbers

Primary: Counting
Early Addition and Subtraction
Tens and Ones

Upper: 2 digit addition and subtraction
Rounding
Even & Odd Numbers
Multiplication & Division
Fraction Concepts

The Assessments are color coded, to make it easier. Sections in green are assessed to all students, pink primary, or before students have developed place value concepts, and blue assesses children with at least some place value understanding.

The length of the average assessment is about 30 minutes, and might be faster with younger children and longer with older. In a k-6 setting, I generally do not have students who are able to answer many of the fraction questions, at least until my 5th or 6th graders are ready to exit from my services.

<u>What you will find inside:</u>

➢ **Data Collection sheets**

✓ **Score sheet for the Developmental Math Assessment**

The score sheet allows the assessor to document how the child does on all sections of the assessment. Only the sections assessed are filled out. The developmental level of the student is recorded on the top of the assessment. Scoring directions are given in the assessment directions. In most sections, a score of 0-3 will be used.

✓ **Individual Assessment Summary Sheet**

This form can be filled out for each student assessed and provided to either the teacher or parent, or can be placed in the child's CUM folder. There is a spot for indicating the developmental phase of the student as well as that of his same-age peers. The results for the individual sections are recorded as well as a written narrative of how the child performed during the assessment. This information is helpful in justifying the need for support for the child, and comparing child performance from assessment to assessment.

✓ **Classroom Assessment Spreadsheet**

This spreadsheet allows the assessor to record the scores from the individual subtests, as well as the developmental phase, for all the students in the class or on her caseload.

✓ **Developmental Phase Data spreadsheet**

This can be useful in either determining the phase of the student, or to record observed mathematical behaviors. It also serves as a reminder of what might need to be addressed with that individual.

✓ **Worksheet for the Upper version of the assessment**

The upper version of the assessment requires a worksheet to record the thinking of the student. One worksheet should be printed for each student. This is for the teacher's purposes and not to be used by the student.

➢ **Assessment directions & Intervention Strategies**

The directions for each subtest are included in the tabbed section of the manual followed by intervention strategies / activities that can be used with children who are struggling in that particular area to help build conceptual understanding.

➢ **Task cards**

Task cards are provided for sub tests that require a problem or diagram to be shown to the student.

General Assessment Directions

Using the Developmental Assessment Score sheet, start by administering the subtests in green: Combination of numbers to ten, Ten and some more, and Reading & Comparing Numbers. Then administer either the primary or upper version of the test, depending on the age and developmental level of the child. Choose primary for those who have not yet developed an understanding of place value.

Students will be able to do some problems from each subtest, however, within the subtest, a child may only be able to do the first few question since each one gets progressively more difficult. Using the scores gathered on the score sheet, fill out an Individual Assessment Summary sheet for each student, if desired, and record the data on the Classroom Assessment Spreadsheet and Developmental Phase Worksheet. There are some indicators on the Developmental Phase Worksheet that are not assessed on this test. It serves as a reminder of other items that are included in the phase so the instructor can remember to work with the child in those areas as well.

Note: *The forms contained are available as a fillable .pdf document. If interested, please visit* www.developmentalmathassessment.com *for details.*

Providing Intervention

When considering the intervention techniques I use for students, I use brain research to help guide me. I know that specific areas in the left hemisphere used in counting, calculating, and using basic arithmetic number symbols are located mostly in regions of the left parietal lobe and motor cortex. Areas in the prefrontal cortex are used in analyzing a problem and retrieval of facts. Regions in the right parietal lobe are used in spatial reasoning and visual-spatial tasks, like being able to generate a mental number line, and estimating.

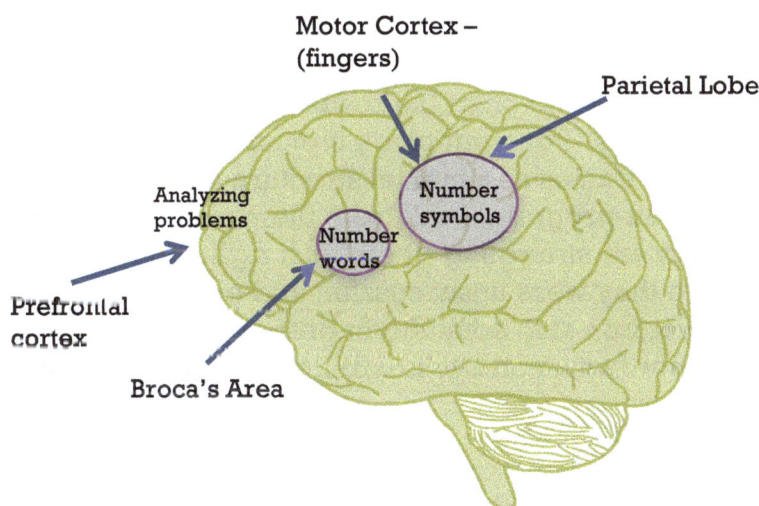

Motor Cortex –
(fingers)

Parietal Lobe

Analyzing
problems

Number
symbols

Number
words

Prefrontal
cortex

Broca's Area

Source: *How the Brain Learns Mathematics,* by David Sousa

What I have found, through experience, is that if students are somewhat lacking in the visual-spatial areas on the right hemisphere, they are usually able to compensate with the computing ability of the left. They may not be conceptualizing the math, but they are able to get by, since traditional school culture tends to put more value on computation rather than understanding. Computation is a predominately left-brained skill. Mathematics, on the other hand, is much more encompassing. It is about understanding the relationships between numbers and objects and the patterns that create the universe.

Most of the children that end up in my program tend to be holistic, kinesthetic, and visual thinkers and have significant weaknesses in areas on the left hemisphere that effect their ability to compute or recall basic facts. They may equally have difficulty in reading since decoding and phonemic awareness are also located on the left side. They may have labels such as dyslexia and/or dyscalculia. Their right side areas are generally functioning properly or may even be well above average. If these children are not in an environment that embraces understanding and conceptual thinking, they will have limited access to understanding mathematics, even though they may very well be destined to be great mathematical thinkers.

We have number sense because numbers have meaning to us. When providing intervention, it is important to help develop the child's number sense through concrete and visual means. Although they may be visual thinkers, these children may not have had a chance to make sense of our number system, nor have developed mental structures for thinking about number. I start by providing inputs for students to expand their ability to recognize quantities and to begin to create mental structures. I use dot cards, color tiles, and ten frames for this. I use the visual to help the students make sense of the numbers. I have had students whom I have described as having a "sea of numbers in their heads." I had to help them create organizing systems in order to think and deal with numbers.

Putting it together

Once I conduct the assessment and chart the results, I will usually create a program that incorporates several elements. For example, a typical student struggles with number combination, the use of ten frames as a mental structure and the basic understanding of base ten. Therefore, in their program, I will have them practicing the number combination at which they are struggling (their "just right number"), working with ten frames at the point where they were unsuccessful, and engaging in place value activities. I might also have them work on skip counting in preparation for later work. This way the child is working on number facts, number flexibility, and building understanding around the place value system.

Throughout this manual, I will describe some of the ways I introduce concepts that move along the concrete----abstract continuum. In this way, we are helping students access the mathematics through using their right hemisphere (concrete / visual) so they can make sense of it and make connections to the left hemisphere (symbolic). It is through the connections between all the areas of the brain involved in mathematics that make an individual a proficient mathematical thinker and problem solver.

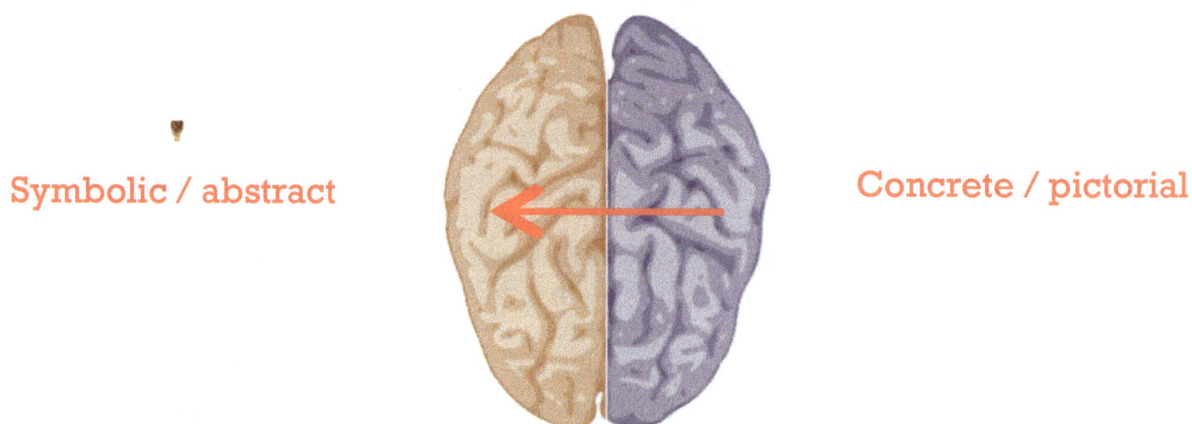

Symbolic / abstract   Concrete / pictorial

Developmental Phases

Emergent Phase	Matching Phase	Quantifying Phase
Counting • Know numbers signify quantity • Rote count to 10 with words in order • Identify by sight 1–5 objects • Begin to write numerals • Know numerals are different from letters • Know more, less, and same and that a change results in either bigger or smaller (but not know by how much)	**Counting** • Rote count to double digits • Count lots of collections • Know one-to-one correspondence — Keep track through moving, touching, and then pointing — Get a particular quantity — Remember how many after counting • Count all or use direct modeling in problem solving • Know the cardinality principle (last number counted stands for total amount; it is not just a label) • Know what things can be discretely counted versus continually counted **Number Relationships** • Know more, less, and same • Figure out how many more or less or how many to make the same amount • Know one more and one less without counting • Know spatial relationships and beginning estimation • Relate one number to another (when changing numbers) — Know if more need to be added or take away — Put more or remove extras	**Counting** • Count on, count back • Skip count (in groups) — Realize repeated addition or subtraction results in same total as counting by ones • Know ordinal numbers **Number Relationships** • Develop a sense of quantity and reasonableness • Understand part–part–whole relationships — • Know numbers are within numbers — Understand number combinations • Combine by using relationships, using doubles and near doubles, and just knowing • Use benchmarks of 5 and 10 • Know how many to add or subtract (to change a number) • Relate one number to another — Know how many to add or subtract • Have conservation of number • Write number sentences (equations) • Understand greater than, less than, and equal to • Understand that when dealing all groups are the same • Use concrete materials to model tens and ones and to add tens

Partitioning Phase	Factoring Phase	Operating Phase
Counting • Count by tens starting with any number (e.g., 34) • Skip count while keeping track of number of groups counted (double counting) **Number Relationships** • Use part–part–whole relationships without seeing objects • Know any number is made with other numbers • Understand and use inverse operations for addition and subtraction • Compare whole numbers using patterns that do not concretely represent amounts (e.g., 100 chart) **Numbers as Tens and Ones** • Recognize numbers as tens and ones • Combine and separate tens and ones — Tell how many to make the next 10— Add by making tens— Subtract by breaking apart tens and restructuring remainder into tens and ones • Know 10 more or less for any two-digit number • Know place value for two-digit numbers • Know and use expanded notation Fractional Understanding • Believe equal halves can look different • Divide numbers into fractions • Know 1/3 is greater than 1 / 4	**Numbers as Hundreds, Tens, and Ones** • Know place value for three-digit numbers and larger • Read, write, and say whole numbers beyond thousands **Multiplication and Division** • Use arrays to represent multiplication • Understand inverse operation for multiplication and division • Decompose and recompose factors without changing quantity • Understand and use the commutative property of multiplication • Think additively and multiplicatively • Understand different models for division • Use other language to interpret x and ÷ signs (groups of, shared by, etc.) **Fractional Understanding** • Represent fractions with models and pictures and compare their like and unlike denominators • Split fractions and decimals into whole and parts • Relate fractions to division	**Counting, Place Value, and Number Relationships** • Count by tenths, hundredths, and thousandths over the whole • Use understanding of relationships between successive places to order decimal numbers **Multiplication and Division** • Make multiplicative comparisons and deal with proportional situations • Understand that when multiplying by a factor less than one, the product is smaller • Understand that when dividing by a divisor less than one, the quotient is larger **Fractional Understanding** • See any number can be thought of as a unit that can be repeated or split up a number of times • Represent common and decimal fractions on a number line • Partition decimal numbers • Understand two fractions are being compared to same whole • Compose and decompose fractions visually or mentally • Write number sentences (equations) for multiplication and division of whole numbers, fractions, and decimals

Compiled from Western Australia's Department of Education and Training (Willis et al 2006) and Kathy Richardson's *Developing Number Concepts (1999a-c)*. Printed in Bahr/de Garcia's *Elementary Mathematics is Anything but Elementary: Contents and Methods from a Developmental Perspective*, 1E. ©2010 South-Western, a part of Cengage Learning, Inc. Reproduced by permission. www.cenage.com/permissions.

Developmental Math Assessment
Score sheet

K-2	All	3-up

Student Name: Teacher Name: Date:

Developmental Level: Matching___   Quantifying____   Partitioning____   Factoring___   Operating_____

_____Combination of Numbers to Ten

4	5	6	7	8	9	10

_____Ten and Some More

_____Recognizes 1-5 on ten frame   _____Can make a 10 with 9+ using 1 visual

_____Recognizes 1-10 on ten frame   _____Can make a 10 with 9+ mentally

_____Knows 10 and some More (addition)   _____Can make a 10 with 8+ visually (1 or 2)

_____Knows 10 and some More (subtraction)   _____Can make a 10 with 8+ mentally

_____Can make a 10 with 9+ using 2 visuals   _____Generalizes to larger numbers

Comments:

3=Confident   2=Needs practice   1=Needs instruction   0=Frustration level

_____Reading & Comparing Numbers

Reading Numbers

2 digit____   3 digit_____   4 digit_____   5 digit_____   6 digit_____   7 digit_____   8 digit_____   9 digit_____   10 digit _____

tenths____   hundredths_____   thousandths_____

Comparing Numbers

Compares 1 digit numbers_____   Compares 2 digit_____   Compares 3 digit _____   Compares 4 digit_____

compares tenths_____   Compares hundredths_____   Compares thousandths_____

_____Counting

Counting a set quantity   Counting out a quantity from larger pile

     Number size: _____   Number Size_____

One More: Number posed_____   Y___N____   Number Posed_____   Y__N____   Number Posed_____   Y___N____

Once Less: Number posed_____   Y___N____   Number Posed_____   Y__N____   Number Posed_____   Y___N____

Rote counts to: _____   Counts back from: 100_____30   15_____   10_____   <10_____

_____Early Addition and Subtraction

Seeing Groups

   Recognition of small groups to 5_____

Identify and combining parts:

Total dots_____ score_____	Total dots_____ score_____
Total dots_____ score_____	Total dots_____ score_____
Total dots_____ score___	Total dots_____ score_____

Going from one number to another

   Number range _____ to ____;   ____ to ___;   ____ to ____

   Tells if more or less is needed: More_____ Less____

   Tells how many more_____   Tells how many less _____

Number Relationships

+1____   -1____   +2____   -2____   +3____   -3_____

_____Tens and Ones

_____Gives reasonable estimate

_____ Counts by 2's rote _____Counts by 5's rote ____Demonstrates conservation of number

_____ Number of tens matches estimate _____ Making and counting groups of ten _____Counts by 10's rote

_____ Adds tens _____Subtracts tens _____

Comments:

_____2 digit Addition & Subtraction

_____Identify quantity of tens and ones

_____ Adds multiples of 10 | _____Subtract multiples of 10

_____Adds tens w/o regrouping w/visual | _____Subtracts tens w/o regrouping w/visual

_____Adds tens and ones w/o regrouping w/visual | _____Subtracts tens and ones w/o regrouping w/visual

_____Adds w/regrouping w/o visual | _____Subtracts w/ regrouping w/visual

_____Adds w/o regrouping numerically | _____Subtracts w/o regrouping numerically

_____2 digit ___3 digit | _____2 digit ___3 digit

_____Adds w/ regrouping numerically | _____subtracts w/ regrouping numerically

_____Connects + algorithm to concrete tools/picture | _____Connects - algorithm to concrete tools / picture

_____Creates situation for addition | _____Creates situation for subtraction

_____Uses reason / relationships to mentally compute (+) | _____Uses reason/relationships to mentally compute (-)

Comments:

_____Rounding / Even & Odd

_____Rounds 2 digit to the nearest ten

_____Rounds 3 digit t nearest ten _____ hundred

_____Rounds 4 digit to the nearest ten _____hundred _____thousand

_____Round to the nearest whole_____ tenth_____ hundredth_____ thousandth_____

_____Understands even & odd numbers

Comments:

_____ Multiplication & Division

Multiplication

_____Connect number sentences to 3 x 5 array

_____Connect number sentence to 5 x 5 group

_____Connect number sentence to ten frames

_____Understands the language of multiplication

Skip counts by _____10_____5_____2_____3_____4_____6_____9

_____1 digit factors

_____2 x 1 digit factors

_____2 x 2 digit factors

_____creates situation for multiplication

Division

_____Connect division to 3 x 5 array | _____Solves with 1 digit divisor & dividend

_____Connect division to 5 x 5 group | _____Solves with 2 digit divisor & 1 digit dividend

_____Connects division to ten frames | _____Solves with 2 digit divisor & 2 digit dividend

_____Creates situation for division | _____Order of operations

Comments:

Fraction Concepts

☐ **Identifying half**
Student is able to identify ½ in the following types of objects: ____part of whole ____part of set
____only when whole is split into two pieces _____ Understands that each piece must be of equal size.

☐ **Part-set fractions**
_____says 2/3 _____says 2/5
Student explanation:

☐ **Part-whole fractions**
_____Student is able to label all four pieces
List fractions that the student correctly labels:

☐ **Number Line**
_____Knows ½ _____Knows ¼ , ¾ _____Knows 1/8, 7/8

☐ **Comparing fractions in context**
____Correct ____incorrect ____Self-Correct ____Did not know ___demonstrated conceptual understanding
Student explanation:

☐ **Comparing fractions**
____Correct ____incorrect ____Self-Correct ____Did not know ___demonstrated conceptual understanding
Student explanation:

☐ **Fractions as division**
____Correct ____incorrect ____Self-Correct ____Did not know ___demonstrated conceptual understanding
Student explanation / representation:

☐ **Fractions as a ratio**
____Correct ____incorrect ____Self-Correct ____Did not know ___demonstrated conceptual understanding
Student explanation:

☐ **Fraction Computation – addition in context**
____Correct ____incorrect ____Self-Correct ____Did not know ___demonstrated conceptual understanding
Student explanation:

☐ **Fraction Computation - addition**
____Correct ____incorrect ____Self-Correct ____Did not know ___demonstrated conceptual understanding
Student explanation:

☐ **Fraction Computation - subtraction**
____Correct ____incorrect ____Self-Correct ____Did not know ___demonstrated conceptual understanding
Student explanation:

☐ **Fraction Computation – multiplication in context**
____Correct ____incorrect ____Self-Correct ____Did not know ___demonstrated conceptual understanding
Student explanation:

☐ Fraction Computation - multiplication

    _____Correct   _____incorrect   _____Self-Correct   _____Did not know   ___demonstrated conceptual understanding

Student explanation:

☐ Fraction Computation - division in context

    _____Correct   _____incorrect   _____Self-Correct   _____Did not know   ___demonstrated conceptual understanding

Student explanation:

☐ Fraction Computation - division

    _____Correct   _____incorrect   _____Self-Correct   _____Did not know   ___demonstrated conceptual understanding

Student explanation:

Overall Comments:

Assessment Summary Sheet

Student Name:

Teacher Name:

Date of Assessment:

Developmental Phase: Mark the location within the appropriate phase

Matching	Quantifying	Partitioning	Factoring	Operating
early__mid___late__	early__mid__late___	early__ mid __ late__	early__mid___late__	early___mid___late__

Developmental Phase of same-age peers:

Matching	Quantifying	Partitioning	Factoring	Opcrating
early___mid___late__	early___mid___late__	early___mid___late__	early___ mid___ late___	early___mid___late

	Comments:
_____Combination of number _____ 10 Frames _____ Reading & comparing numbers _____Counting _____Early addition and subtraction _____Tens and ones _____2 digit addition and subtraction _____Rounding _____Multiplication and division _____Fraction Concepts	

N/A = not assessed or not applicable

General Observations during Testing:

Developmental Phase Data Sheet

Student Name	Matching Phase							Quantifying Phase									Partitioning Phase										
	Rote count to double digits	One-to-one correspondence	Get a quantity w/o going over & remember how many after counting	Relates one number to another: knows if more or less is needed	Counts all or uses direct modeling in problem solving	One more/one less w/o counting	Counting on / counting back	Skip Counts	Counts money	Combination # to 10	Combine by relation: doubles, near doubles, just knowing	Relates one number to another: knows how many more is needed	10 + a single digit # / 10 to 20 -10	Making a ten with 9 and 8	Making a ten with larger numbers	Conservation on number	greater than, less than and equal	Recognize #'s as tens and ones	Know and use expanded notation	Counts and adds multiples of 10	Adds with regrouping (C/N)	Subtract with regrouping (C/N)	Connects +- to blocks	Skip count and keep track of groups	Believe equal halves can look different	Know that 1/3 is greater than ¼

Factoring Phase																					Operating Phase								
Know place value for 3+ digit #	Read, write, say whole numbers beyond thousands	Connect mult to groups/array	Understand inverse of mult/div	Think additive and multiplicative	Use language for x and ÷ signs	Understand commutative prop X	Knows multiplication facts	Solves 2x1 digit (repeat + / Trad)	Solves 2x2 digit (Trad / Alt)	Create situation for multiplication	Knows ÷ is inverse of X	Can write ÷ sentence for array	Knows ÷ facts	2 digit divisor 1 digit dividend	2 digit divisor 2 digit dividend	Creates situation for division	Represent fractions and compare their like & unlike denominators	Common decimal/fraction on #line	Split fraction into whole and parts	Relate fractions to division	Count by tenths, hundredths, thousandths over the whole	Able to order decimals	Partition decimals	Compose and decompose fractions visually / mentally	Understand x by factors less than one the answer is smaller	Understand when ÷ by # less than one quotient is larger	2 fractions are being compared to same whole	Any number can be a unit and repeated or split many times	Write equations for x÷ of frac, decimals # whole numbers

Classroom Assessment Spreadsheet

Student Name	Developmental Stage	combination of number	Ten and Some More	Reading & Comparing Numbers	Counting	Early Addition and Subtraction	Tens and Ones	2 digit Addition & Subtraction	Rounding	Even / Odd	Multiplication and Division	Fraction Concepts		Developmental Stage	combination of number	Ten and Some More	Reading & Comparing Numbers	Counting	Early Addition and Subtraction	Tens and Ones	2 digit Addition & Subtraction	Rounding	Even / Odd	Multiplication and Division	Fraction Concepts		Developmental Stage	combination of number	Ten and Some More	Reading & Comparing Numbers	Counting	Early Addition and Subtraction	Tens and Ones	2 digit Addition & Subtraction	Rounding	Even / Odd	Multiplication and Division	Fraction Concepts

Beginning Assessment — **Mid-year Assessment** — **End of Year Assessment**

+Worksheet Assessing those with at least some understanding of tens and ones

| | | Materials: base ten blocks, paper and pencil |

2-digit addition and subtraction

Present this amount in base ten blocks or linking cubes

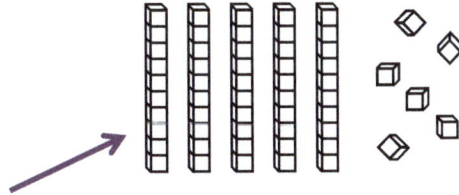

If there are ten cubes in this train, how many cubes are in all?_____

How did you count them?   tens and ones _____   other (describe)_____

How many would there be if we added 20 more (10 more, if too much)? How do you know?	How many would there be if we added 12 more? How do you know?
How man would there be if we added 9 more? How do you know?	How many would there be if we took 20 away? How do you know?
How many would there be if we took 15 away? How do you know?	How many would there be if we took 9 away? How do you know?
Solve: $14 + 12 =$	Solve: $29 - 14 =$
(connect method with linking cubes or picture)	(connect method with linking cubes or picture)

Solve: $$28 + 24 =$$ (connect method with linking cubes or picture)	Solve: $$34 - 16 =$$ (connect method with linking cubes or picture)
Solve: $$281 + 241 =$$ (connect method with linking cubes or picture)	Solve: $$376 - 198 =$$ (connect method with linking cubes or picture)
Mentally Solve: $$1000 - 998 =$$	Mentally Solve: $$99 + 16 =$$
Mentally Solve: $$15 + \underline{\quad} = 200$$	Create your own story which would use addition:
Create your own story which would use subtraction:	

Rounding

What number would this be if you rounded it to the nearest ten? **28**	**42**
What number would this be if you rounded it to the nearest hundred? _____ ....to the nearest ten?_____	**384**
What number would this be if you rounded it to the nearest hundred? _____ ...to the nearest thousand?_____ ....to the nearest ten?_____	**6,139**
What number would this be if you rounded it to the nearest whole number? _____ ....to the nearest ten?_____	**35.6**
What number would this be if you rounded it to the nearest hundredth? _____ ....to the nearest thousandth?_____	**4.0826**

Multiplication & Division

How many squares are there? _____

How did you count them?_____

_____

What number sentences can be written about this picture?_____

What multiplication sentence(s) can be written about this picture?_____

What division sentence(s) can be written about this picture?_____

How many squares are there? _____

How did you count them?_____

_____

What number sentences can be written about this picture?_____

What multiplication sentence(s) can be written about this picture?_____

What division sentence(s) can be written about this picture?_____

How many squares are there? _____

How did you count them?_____

_____

What number sentences can be written about this picture?_____

What multiplication sentence(s) can be written about this picture?_____

What division sentence(s) can be written about this picture?_____

If I have four fives, how many would that be altogether?_____

If I have three fours, how many would that be altogether?_____

If I have six twos, how many would that be altogether?_____

Solve: 6 x 8 =	Solve: 4 x 13 =
Solve: 15 x 10 =	Solve: 60 x 40 =
Solve: 14 x 22 =	Solve: 48 ÷ 6 =
Solve: 180 ÷ 12	Create your own story to show 3 x 4:
Create your own story to show 15 ÷ 3:	Order of operations: 4 + 6 x 2 = 3 x (4 + 7)= 2 x 3 + (3^2 − 2)=

Fractions

1. Identifying Half - Circle each figure that shows ½

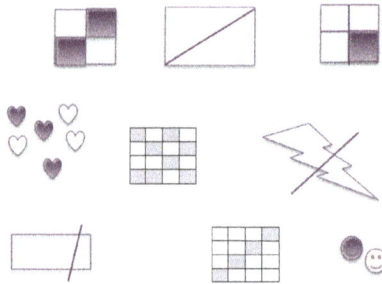

2. Part-Set Fractions

This desert plate has brownies and peanut butter cookies. Kim said that the brownies make up 2/3 of the plate. Sean said that they make up 2/5. Who is correct and how do you know?

3. Part-Whole Fractions

How many fractions can you see in the diagram? Label as many as you can.

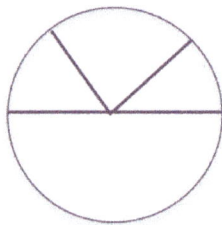

4. Fractions on a Number Line

What is the value of point A? Write the value below.
For each successful response, continue with the next point.

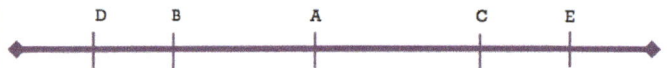

Note: it is important for students to write the fractions themselves since the way they say and write the fractions may differ.

5. Comparing Fractions in Context

The library is ¾ of a mile from my house. The grocery store is 6/8 of a mile from my house. Which one is farther from where I live? Show your thinking

6. Comparing Fractions - Raw Numbers

Show each pair of fractions, one at a time. Ask which fraction is larger, or if they have the same value.

1/2   2/5

3/8   4/5

7. Fractions as Division

Four brothers went home for an after school snack. Their mother prepared 3 sandwiches. Show how these boys are able to split up the sandwiches so each can get the same amount. How much of a sandwich can each child get?

8. Fractions as a Ratio

When making lemonade for their lemonade stand, Joshua and Andrew wanted to make 10 liters of lemonade. For each liter, they used ¼ juice to ¾ water. How much juice and water did they need for the 10 liters?

9. Fractions Computation (addition in context) Susan and Kim want to make brownies for the bake sale. The brownies call for ¾ cup of cocoa. They want to double the recipe and are trying to decide how much cocoa they will need. Susan says they will need 6/8 cup of cocoa and Kim says they need 6/4. Who is correct and how do you know? 	**10. Fractions Computation (addition)** Will the sum be 1, less than 1 or greater than 1? How do you know? Solve to check. $$1/2 \; + \; 2/3$$
11. Fraction Computation (subtraction) Will the difference by ½, less than ½, or greater than ½? How do you know? Solve to check. $$5/8 \; - \; 1/4$$	**12. Fraction Computation (multiplication in context)** Margaret made cupcakes for the bake sale. She made 3 different kinds. For each kind she used 2/3 cup of frosting. How much frosting did she use altogether?
13. Fraction Computation (multiplication) Will the product be larger or smaller than ½ *(point to the first factor)*? How do you know? Solve to check. $$1/2 \; x \; 2/3$$	**14. Fraction Computation (division in context)** Adrian lives 2 ½ miles from school. If every ¼ mile there is a stop sign, how many stops signs are there between his house and school?
15. Fraction Computation (division) Will the quotient be greater or less than 3 *(point to the dividend)*? How do you know? $$3 \; \div \; 1/2$$	

Combinations of Numbers to Ten

Assessment

To assess the student to see if they know combinations of numbers to ten, start with 4 color tiles on the table. Use all the same color and one that is in contrast to the table. For example, I usually like red, but lately I have been assessing on a red table, so I have been using yellow.

Ask how many tiles the child sees, to determine the total quantity. Then put your hand over a few and ask how many tiles are under your hand. Notice how he is figuring them out and do all the different combinations for that number. Make sure it is not obvious as to how many you are hiding. For example, put your hand over the entire amount, then move some away to reveal the rest. If the child thinks it is a magic trick and starts guessing unreasonable numbers, you must stop. If he is counting up, counting back, using fingers, or any other way to figure it out without knowing, then you need to stop and not continue to the next number. This will be his "just right number." Use that number with the activities and games on the following pages until mastery is reached, then move to the next number until he has mastered all combinations of all numbers to 10. Teachers of older students will want to use this assessment to determine where to begin with practicing the basic facts.

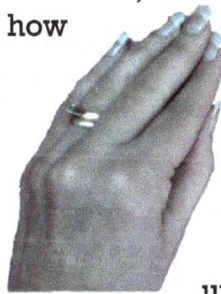

Hiding Games

Use Objects, such as color tiles, pennies, beans, etc., and lay them out on the table, using the number that the child is learning to master. Cover up some of the objects and have the child guess how many are hiding. Keep hiding different amounts and notice how the child is figuring them out. Is he eyeballing and counting what he sees and counting up to the desired number? Is he starting with the number and counting backwards using the tiles he can see? Is he using fingers to figure out how many are left? Does he just know? It is important that the child knows the combinations, so do not consider it mastered until you are convinced that he just knows, or he is able to use a strategy so fast that it is hard to tell.

For independent activities, envelopes or paper bags can be used. Children put the total number of cubes, pennies, popsicle sticks, etc., into a bag, grab a handful without looking in the bag and guess how many are remaining based on how many are in his hand.

Part-Part-Whole Dot Card

Cut cardstock sheets long-ways in half. Then create fold about 1/3 of the way

Put red dot circles inside so that some are visible when folded and the others are not

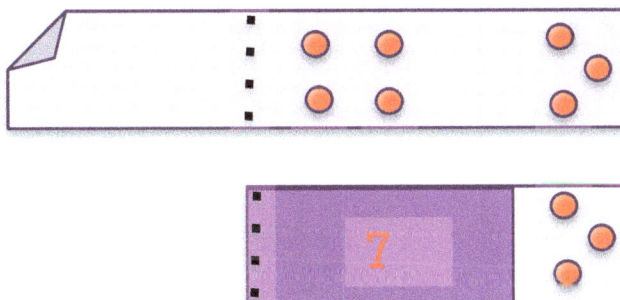

Students guess how many dots are hiding.

Note: If you start with the number "4," and you include zero in the combinations, it will take about 420 dots to make all the combinations through ten.

Fun note: (As I was figuring it out, I noticed that I needed to multiply the number times the next number to figure out the total dots used for just that number: 4 x 5 = 20 dots needed for all the combinations of 4)

Other practice tips

Did you know that on a 6-sided die, the opposite sides total 7? Have the student roll a die and guess the number that is on the bottom. This helps learn the combinations of 7 quite quickly.

I also have students make dot flash cards where they put the number of dots on either side to total the number they are working on, such as 3 dots on one side and 4 on the other to total 7.

Addition and Subtraction Facts

Use Flash cards, oral quizzes, or online addition/subtraction games, such as Timez Attack™ (from Bigbrainz.com): addition and subtraction, to reinforce combinations that your students are learning or have already learned. Cut out the cards below of the number they are working on.

4 −1	4 −2	4 −3	4 −4	4 −0
1 +3	3 +1	4 +0	0 +4	2 +2
5 −1	5 −2	5 −3	5 −4	5 −5
5 −0	4 +1	3 +2	2 +3	1 +4

0 +5	6 -1	6 -2	6 -3	6 -4
6 -5	6 -6	6 -0	5 +1	4 +2
3 +3	2 +4	1 +5	6 +0	0 +6
7 -0	7 -1	7 -2	7 -3	7 -4

7 -5	7 -6	7 -7	0 +7	1 +6
2 +5	3 +4	4 +3	5 +2	6 +1
7 +0	8 -0	8 -7	8 -6	8 -5
8 -6	8 -7	8 -8	8 +0	7 +1

6 +2	5 +3	4 +4	3 +5	2 +6
1 +7	0 +8	9 -0	9 -1	9 -2
9 -3	9 -4	9 -5	9 -6	9 -7
9 -8	9 -9	9 +0	8 +1	7 +2

6 +3	5 +4	4 +5	3 +6	2 +7
1 +8	0 +9	10 + 0	9 +1	8 +2
7 +3	6 +4	5 +5	4 +6	3 +7
2 +8	1 +9	0 +10	10 - 0	10 - 1

10 - 2	10 - 3	10 - 4	10 - 5	10 - 6
10 - 7	10 - 8	10 - 9	10 - 10	

✛Ten and Some More

Materials: One set of ten frames

Assessment

One of the goals with using the ten frames to create a mental tool is for children to have a way to generalize a rule for adding numbers, such as 9, by using "making a ten" strategy. The following assessment steps will determine the level of student readiness for using this strategy.

First, students must be able to recognize the dots on a ten-frame. Without this, students will not be able to use the ten-frame as a mental structure at all. Most children are able to instantly recognize 1-4 without counting (this is called subitizing), even without experience of seeing the frames. In fact, it a red flag if there are children in kindergarten and first grade and are unable to identify 4 objects without counting.

1) Recognizes 1-5; 6-10 on a ten frame:
    If children are first grade or younger, I generally present the 1-5 dots first (out of order). If they are able to identify without counting, I will then present 6-10 (out of order). If students have to count the dots and are unfamiliar with the ten frames, then they will most likely not be able to use them at the moment to make a ten, but I will still continue the assessment just in case they are visual enough that they are learning it on the spot. Give a point on the score sheet if they know all without counting. You might want to note which ones they count for future reference. You can give a ½ a point if they know most, but not all.

2) Knows 10 and some more (addition)
    Ask the student if s/he knows what 10+6 is. You may want to ask a few problems like this. If s/he knows without counting, give a point. *If s/he does not know this, then stop the test. S/he will not be able to use a making 10 strategy.*

3) Knows 10 and some more (subtraction)
    Ask the student if s/he knows what 15-5 is, then 17-10, or similar problems. The point is you are trying to see if they know that the number is made of ten and left-overs and they know if they take away ten they are left with the left-overs. If they do not know, you might try placing the frames with ten and five dots to see if they can do it visually, just for your information. If they know both ways mentally, they get a point. Give ½ a point if they can only do one of the two types of problems. Note which, if any, they were able to solve. If they received a point for number 2 (above), but not on this one, you still may proceed. Many students cannot do the subtraction, but are able to use make a 10 strategy, *continue to step 4.*

4) <u>Can make a 10 when adding 9+ a number</u>
Show the frame with 9 dots and the one with 4 dots
below. Say, "how many there would be you had 9 and
4 more?" Notice how s/he solves the problem and ask
the student to explain his or her thinking. They may
count all, count on, or mentally move the dot into the
empty space. Each will tell you something about his
or her developmental level. If they count on, you may
want to ask if there is a way the dots can help. Sometimes they never thought
about it before, but can figure out a way on the spot. If the student can articulate
somehow that they are moving the dot to make a ten, or they just know then they
get a point and you can move on. ***If not then you must stop the test.***

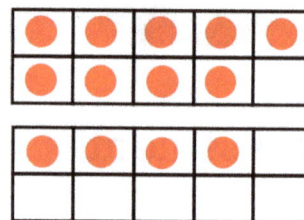

5) <u>Can make a 10 with 9+ using 1 visual</u>
Show the frame with 9 dots and say, "if you had 9 and
6 more, how many would you have altogether?"
Notice if the child still uses a making 10 strategy (or
just knows), or if they revert to counting on, either
with the dots or their fingers. Give the student a point if s/he is still able to
accurately tell the sum by just knowing or making a ten. ***If they revert to
counting, then you must stop the test.***

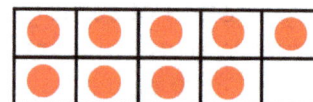

6) <u>Can make a 10 with 9+ mentally</u>
Simply ask the student, "if I had 9 and 5 more, how many would I have
altogether?" Notice if the child still uses a making 10 strategy (or just knows), or
if s/he reverts to counting on, either with the dots or with fingers. Give the
student a point if s/he is still able to accurately tell the sum by just knowing or
making a ten. ***If s/he reverts to counting, then you must stop the test.***

7) <u>Can make a 10 with 8+ visually</u>
Present a frame with 8 dots and one with 4 dots below and ask, "if you had 8 and
4 more, how many would you have altogether?" See if
the student is able to still use the making a ten strategy
when only 8 dots are used. Repeat the question with a
different number, but this time do not show the second
frame. Note if there is any difference when they are only
able to see the frame with 8 dots. Give the student a point
if he/she is able to make a ten with either one or both
visuals or just know the answer. ***If s/he reverts to counting, then you must stop
the test.***

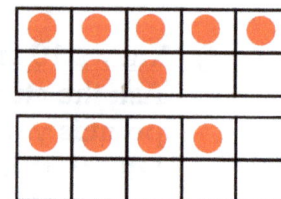

8) <u>Can make a 10 with 8+ mentally</u>
Simply ask the student, "if you had 8 and 7 more, how many would you have
altogether?" Notice if the child still uses a making 10 strategy (or just knows), or
if s/he reverts to counting on, either with the dots or with fingers. Give the
student a point if s/he is still able to accurately tell the sum by just knowing or
making a ten. ***If s/he reverts to counting, then you must stop the test.***

9) Generalizes to larger numbers

Since the answer in the task above would have been established to be 15, then ask the student, "so if 8 and 7 are 15, then how much would 18 and 7 be?" Notice how s/he is able to deal with the extra ten. Does s/he see it as an extra ten or does s/he think of 18 as a completely different number, thus reverting back to counting on? Give the student a point if s/he is able to accurately tell you the sum with an explanation that somehow involves the extra ten.

End of Test

Identifying the Dots

For children who need to first learn to recognize the quantities of the frames, provide a lot of exposure. Kindergarten and first grade classes may want to use the ten frames to count the days of school. Older students that are struggling may have the different amounts posted on a wall and while jumping on a trampoline say how many dots are on each frame. After several attempts at counting, children will usually remember. In whole group activities engage students in problem solving with the ten frames, such as asking how many dots they see and how many more would make ten.

Because recognition of small quantities without counting (subitizing) develops in infancy, if the child struggles with recognizing 4 or 5, help them by also providing exposure to dots in different patterns and in random placements. Broadening a child's ability for subitizing positively effects their ability for counting and overall number sense.

Days of School

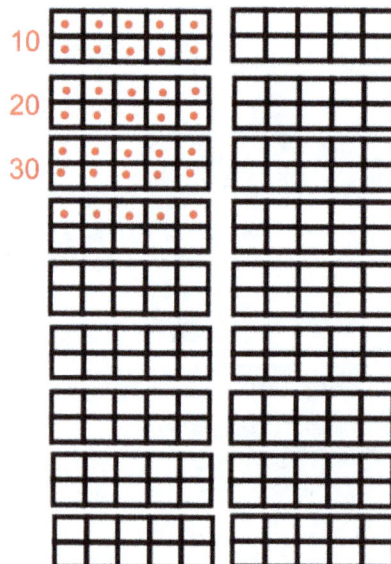

10 and Some More addition

When helping students, who are ready, learn to add ten to a number, the easiest way I have found is through ten frames. Show a frame of 10 dots and a frame with either 0-9 dots. Ask how many they see and record that number sentence on a piece of paper. Change the second addend and have them solve, again recording the number sentence on the paper. Do this until you have gone through all the numbers 0-9, then have the child look at the paper and see if they notice any patterns. If you presented them in order, they may notice that the numbers were increasing by one. They also may notice that there is a 10 in each problem. You may need to direct their attention to the relationship of the numbers on either side of the equation. Children can usually notice that they see a 4, for example, on both sides, and a 1 as well, but not a zero. Ask them to see if that happens all the time. Second graders who haven't learned this will generally have it after one session.

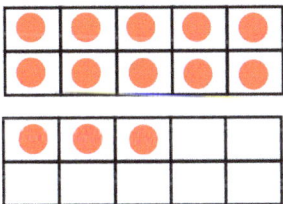

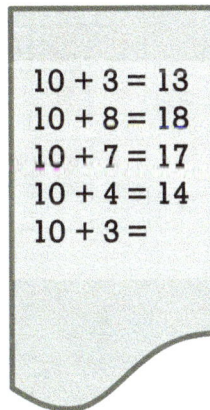

$$10 + 3 = 13$$
$$10 + 8 = 18$$
$$10 + 7 = 17$$
$$10 + 4 = 14$$
$$10 + 3 =$$

Younger students, or those who are not ready, just need a lot of opportunity to build numbers from 11-20 on ten frames placing cubes into the squares and filling them up to match a given number. They can even record the number sentence to match.

16

$$10 + 6 = 16$$

10 and Some More Subtraction

Students who are unable to subtract 10 from a number 11-19, generally are not thinking in tens and ones. The way to help them to do this, often in just one sitting, is again through the ten frames. Simply build a number from 11-19 using the ten frames, for example, 17. Then say, "if I have 17 and I take away 7, how many would I have left?" At the same time, remove the frame with 7 dots so they can see the ten is still there. Put the 7 back and now say, "if I have 17 and I take away 10, how many would I have left?" Again they see that the other frame remains. Do this over and over with different numbers then stop removing the cards to see if the child can mentally remove the card. As the confidence increases, try doing the same problems without the frames present to see if they have generalized the idea of removing the ten or ones. If not, continue with the visual.

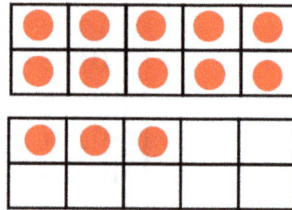

Making a Ten

Making a ten is a powerful addition strategy and involves number relationships and the ability to break apart numbers, which is why it is important that students are comfortable with their combinations of numbers to ten. For example, when adding 28 + 35, a flexible thinker will split the 5 into 2 and 3 and give the two to the 28 to make 30, changing the problem to 30 + 33, which is much more manageable.

To teach children how to make a ten, present a frame with 9 dots and a frame with 1-8 dots on it (start with less than 5). Put a red color counter in the next empty square so that the frame with 3 dots now looks like 4. Ask the child how many 9 and 4 more are. If s/he counts, ask if there is any way that the dots can help. If you are doing this activity as part of a number talk, or whole class discussion, then there will surely be at least one student that will solve by moving the dot either physically or mentally. If you are doing this in a 1-1 setting and the student is unable to come up with that way, prompt him/her with questions such as, "is there any way that the dots can help you," or "is there any way that thinking about enough to spark an idea tens can help?" Usually this is of moving the chip.

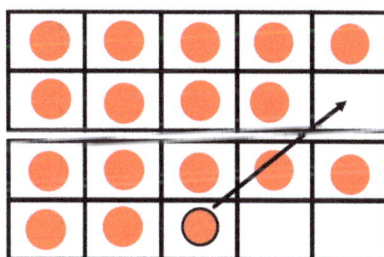

Depending on age, students may need a lot of practice with moving the chip. Eventually they will stop moving it and responding by imagining that they moved it. Once they no longer need to physically move the chip, you can start using the ten frames at face value and not add an additional dot. Whether in a whole group in first grade, or 1-1 in third grade, once students appear confident, you can test the waters by only showing the 9 and not providing the other frame and just say the number or show a numeral. If the students revert back to counting on their fingers, then they need more practice with the visuals. If they are able to mentally solve with just seeing the 9 frame, you can test the waters by not showing either and just orally, or by showing the two numerals, present the problem.

Teachers are often surprised when they first replace the 9 frame with an 8 frame to see if students can use a similar strategy of taking away from the second number. It is really interesting that children usually treat that as a completely separate skill and do not generalize what they were doing with nine. Do not be surprised if this happens to you too. Do the same sequence as with the nine frame, but now using the eight. First with both frames, and dots if necessary, then with one frame, and then removing the frame and using only the numerals.

By the time they have made a generalization for 8+ a number, coming up with one for 7+ a number will be less difficult.

Generalizing to larger numbers

We now need to help the students to generalize these rules to larger numbers, but adding ten to the mix complicates things, in their mind. Simply present a problem, such as 9+ 6 with the frames. Then add a ten frame on top of the nine and say, "now, what if we had 19 + 6." They can see that there is an extra ten, so help them, if necessary, know how to deal with that ten in their answer. Increase the first addend by multiples of ten, such as 29 and 39, while keeping the second addend a single digit. Then go back to 19 and add a ten to the second addend, making the problem 19 + 16, increasing until the student is able to solve mentally problems that involve larger numbers, such as 49 + 36.

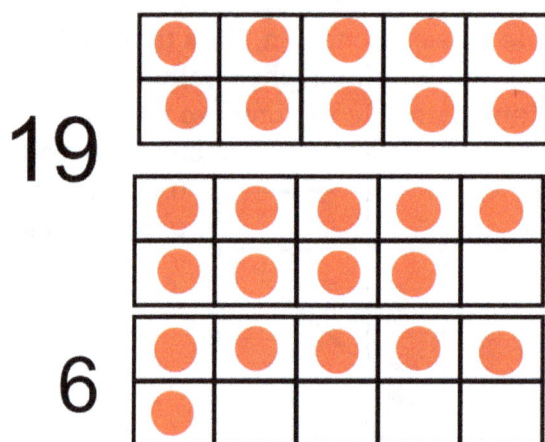

⁺Reading Numbers

Materials: Number cards

Assessment

Show the student numbers of the following number sizes, one at a time. Provide as many numbers in a specific size until you can tell with confidence how well the child performs within that size. In numbers that are 3 digits and larger, provide one with a zero in the middle to see how they handle reading such a number. In numbers with decimals, provide examples with and without whole numbers as well as those with zeros in the number to the right of the decimal. End assessment when student is no longer able to read the numbers.

Number Size

2 digit numbers	6 digit numbers	10 digit numbers
3 digit numbers	7 digit numbers	Numbers with tenths
4 digit numbers	8 digit numbers	Numbers with hundredths
5 digit numbers	9 digit numbers	Numbers with thousandths

Scoring

(3) Student is confident with reading numbers within the specified number size

(2) Student can read some numbers within the number size, but perhaps not tricky ones, such as those with zeros in the middle.

(1) Student can read a limited amount of numbers within the number range

(0) Student cannot read any numbers within the number range

The information from this assessment will allow the teacher to know with which number range the student needs to work on reading numbers. This test does not assess what the child understands about quantity.

+ Comparing Numbers

Materials: Number cards, straws

Assessment

The information from this assessment will allow the teacher to know with which number range the student needs to work on comparing numbers. Provide two different number cards for the students to compare and tell which is largest and smallest. Provide as many numbers in a specific size until you can tell with confidence how well the child performs within that size. Provide some numbers that start with the same number and some that do not. You can also have the child place the correct equality/inequality sign between the two numbers using two small straws taped together at the vertex for a greater than and less than sign and two taped as an equal sign. This way you get information whether the child knows which number is greater as well as how to notate it with the appropriate symbol. End assessment when student is no longer able to compare the numbers.

Number sizes

One digit numbers	Tenths
Two digit numbers	Hundredths
Three digit numbers	Thousandths
Four digit numbers	

Scoring

(3) Student is able to determine which number is larger and smaller within a specific set of numbers, as well as notate it with the appropriate sign.

(2) Student is able to determine which is larger or smaller, but not notate it appropriately

(1) Student is inconsistent when comparing two number

(0) Unable to accurately compare two numbers within the specified number size

Materials: Number cards, linking cubes

Building numbers

If children do not know how to compare numerals, they need to start by building two amounts with cubes and put them side-by-side. The taller tower is the larger amount. Most likely children have not developed enough language around comparison: more than, less than, greater than, etc. Helping them to develop the appropriate language as they are building will be important if helping them understand the concepts around comparing numbers.

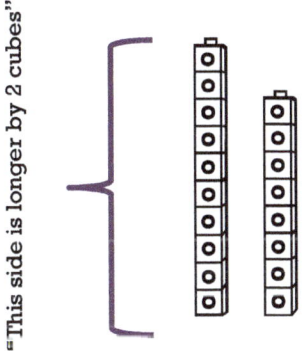

"This side is longer by 2 cubes"

Equality and Inequality signs

Materials: Number cards, cubes, popsicle sticks or pencils

$$= \quad < \quad >$$

The symbolic notation of equality is actually much easier to teach than people think. Most of us all grew up trying to rationalize the greater than and less than signs by the mystical alligator that eats the larger number. I am here to officially announce that alligators do not eat numbers!

How then do we get children to rationalize the symbols? Easy. First, have them build two trains of 4 cubes and stack them next to each other, lying down on the table in front of them.

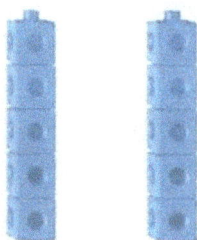

Have them place a pencil, or stick, on the top and one on the bottom of the stacks.

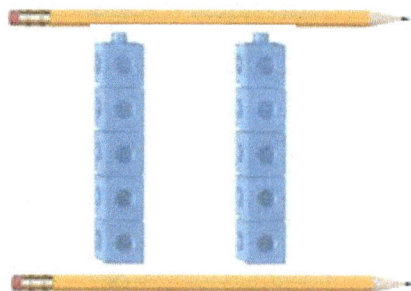

Then have students remove the cubes and ask what symbol that looks like (equal sign). Have the then replace one of the trains and to the other, add another 4 cubes and place it back on the desk and adjust the pencils so they are bordering the top and bottom.

46

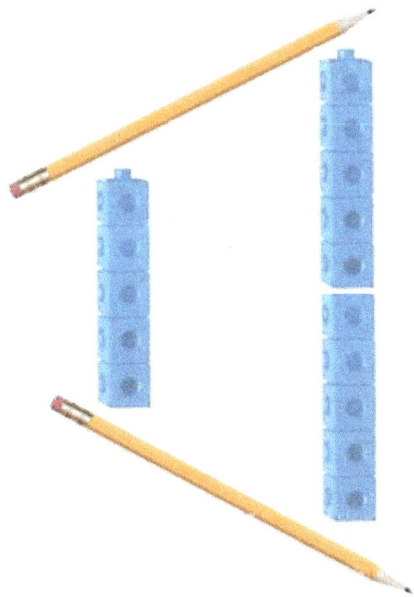

Ask students what happened to their symbol. If they have seen it before, they will say it made a "greater than" sign. If not, they will say it got bigger on one side. Ask why so they can rationalize that the wider side is where there are the most cubes. Do this for a couple more problems, somewhere the opening is on the left as well.

Now, with number cards from a 100 chart, show 2 cards next to each other and ask how they would build that with their pencil. That can be faded to using the straws that are made into the greater than, less than, and equal signs in between the two numbers.

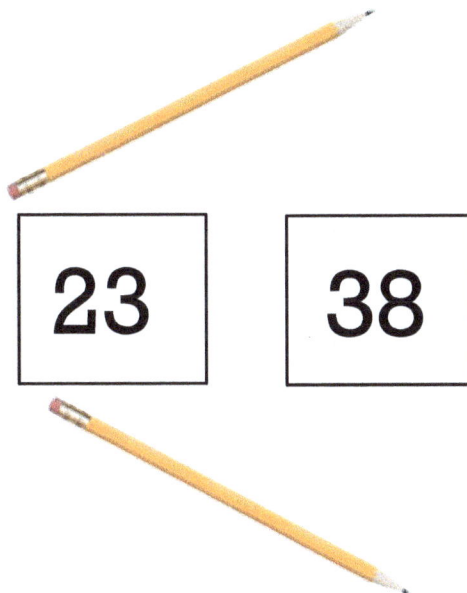

Using Understanding of Place Value to Compare Numbers

Using a place value mat, student builds two numbers and sets one on top of the other. Ask him which is more. After several turns, they will start to look at the amount of the first place. So even though there may be more ones, they will realize that it does not matter until you provide numbers that have the same number of tens. The same can be done when working with decimal numbers as well.

Cube
represents
one whole

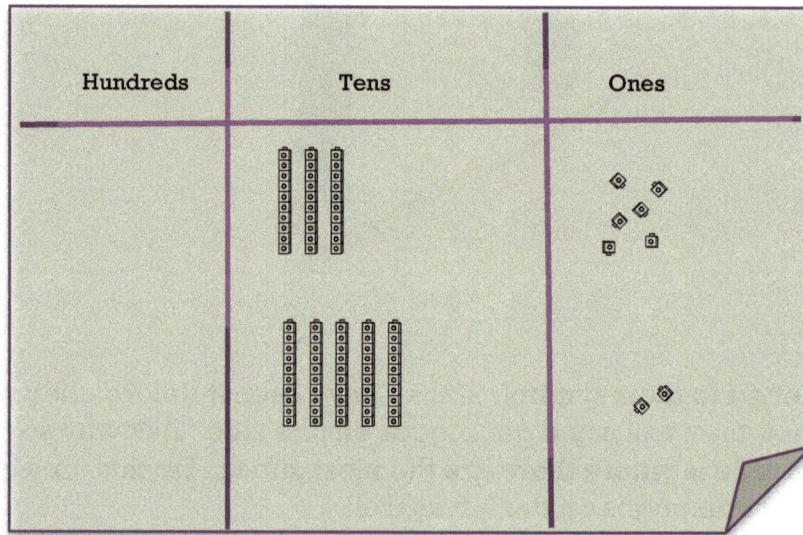

Hundreds	Tens	Ones

36

52

Flat
represents
one whole

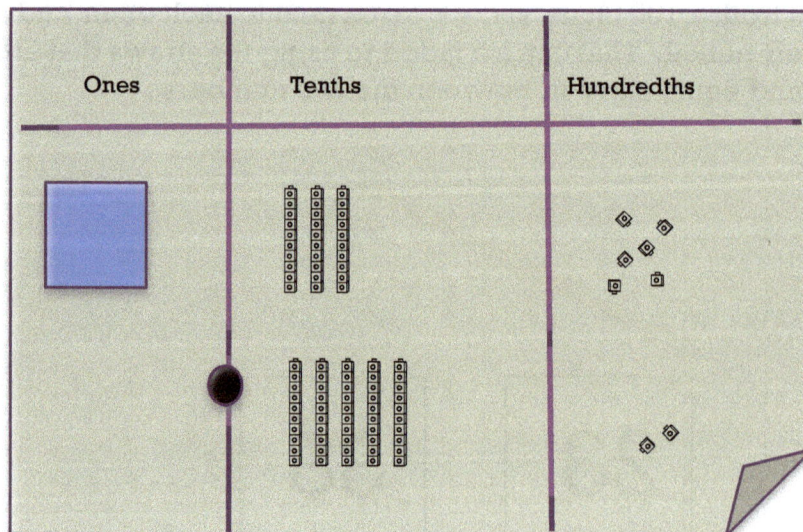

Ones	Tenths	Hundredths

1.36

.52

48

⁺Counting

Materials: color tiles or other counters

Assessment

1) Estimate and check
Provide a pile of counters in front of the child and ask him to guess how many there are. Ask student to check and see.

Note: The size of counters should reflect the number range of the student. If you are unsure, start with about 12. If the child counts with confidence and accuracy, then provide a pile of about 32. If not, reduce to a number under 10. Note how the quantity affects the counting strategy of the child.

> Note the number posed and rate:
> (3) Independent Level - Confident & accurate
> (2) Needs Practice
> (1) Instructional Level
> (0) Frustrational Level

2) Counting out a specified amount
Provide a large pile of counters and ask the student to count out a small pile of a specified amount. Use the number size of the independent level first, then move up from there if s/he is successful. Notice if the student remembers to stop with the number you provide.

> Note the number posed and rate:
> (3) Independent Level - Confident & accurate
> (2) Needs Practice – might pass by one or two.
> (1) Instructional Level - might pass but eventually stops and goes back
> (0) Frustrational Level – passes and does not realize or cannot go back

3) Adding and subtracting one
Provide a pile in their Needs Practice level and have the child verify how many there are. Then add another tile and ask how many there are now. Do this a few times to determine how the child is able to add one more. Now, provide a pile that is just over a ten, such as 22 and have the child verify the amount. Take one away and ask how many there are now. Do this crossing back over the ten to see how the student handles it. Try different amounts to see if they have generalized the strategy to larger numbers. Try doing it without counters (i.e, "if I had 30 and took one away…")

Mark the number posed and mark whether or not they were accurate

4) Counting by rote
Ask the child to count as high as they can and note where they stop. Ask the child to count backwards. Pick a number within a range you think s/he can handle.

Counting Practice

<u>Identifying by Sight One to Five Objects</u>

Through counting, small children naturally are able to identify by sight one to three objects (subitizing). As they gain more experience with building and counting quantities, they will be able to identify by sight larger quantities, especially in patterns such as those found on dice. Children working within a very small range of numbers need to be building quantities on the 5 frame to learn to instantly recognize quantities on this tool. Doing so supports number recognition in later phases. Playing a lot of games involving dice is also very supportive of instant recognition of 1-6. I have found it to be a huge red flag for possible problems when children nearing the end of kindergarten cannot recognize 4 without counting.

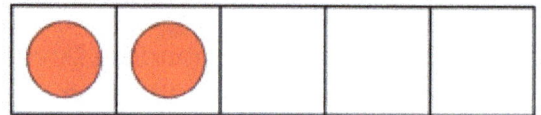

<u>Rote Counting</u>

Rote counting is important, even if children do not yet have a sense of the quantities. They are learning to put words to the quantities they will soon develop, so they can talk about them later. Teachers may count the days of school, the number of seconds it takes the class to settle down, or the number of children present. Teachers must model lots of counting, and children must have lots of opportunities to count. Many kinder teachers may fall into the habit of only counting to 10 or 20, but it is important to model counting beyond 20, as well as counting backwards from 25. This way children learn how to cross the 20. The same goes for when children begin to count 100. They should not stop at 100, but go to at least 120 so they can see how the number pattern continues. This is especially true when skip counting by 10's. Many students I assess will count by 10's to 100, then skip to 200 and 300, because they have not had any modeling or experience of what happens once they get to 100.

22...21...20...19...18...17...

Count Lots

Our youngest students need LOTS of opportunity to count. They need a variety of tools with which to practice and that engages their interest that are authentic and concrete. Many things that the teacher might count may be too abstract (such as the number of seconds it takes to line up). Experienced kindergarten teachers have a myriad of tubs filled with lots of fun things to count, such as dinosaurs, keys, buttons, socks, action figures, and stones.

1-1 correspondence

One-to-one correspondence is not as easy as it appears. In developing one-to-one correspondence, children first learn to keep track of objects through moving, touching, and then pointing or learn other visual ways to keep track, such as bobbing their head up and down, as if counting with their noses. We cannot necessarily impose strategies, such as moving objects while counting, on students until they are ready. When students count by pointing and loose track repeatedly, their teacher might ask whether maybe there is something that they can do to keep track of what they have already counted. Telling them or doing it for them may not support their counting development. Perhaps they need to count in a more concrete way such as a numbered 10 frame, egg carton, paper plates, or number line. Some activities that help with one-to-one correspondence are placing crayons on sheets of paper to see whether there are enough crayons, and passing out supplies to friends in class.

When the number a child is counting is too large, he will usually not stop at the desired number, but pass it, forgetting where to stop. For example, if the teacher asks a student to give her eight cubes, can that student count out cubes from a large pile and stop at 8?

Number size cannot be underestimated. Children may be able to count a number with ease by moving or pointing if the quantity is small enough, but loose track when the amount increases. This idea of number size influencing accuracy does not just apply to counting, but other mathematical skills as well.

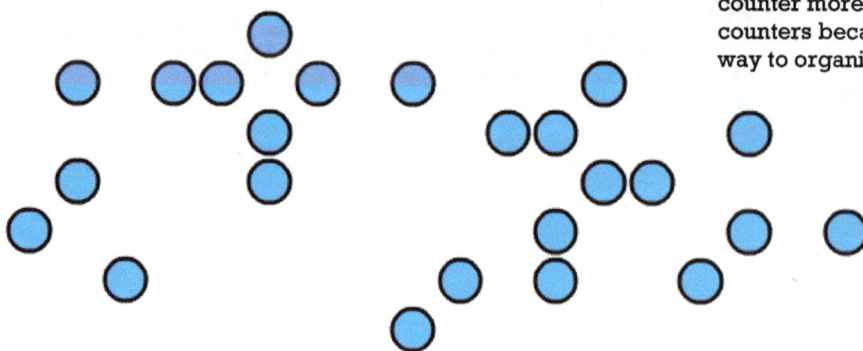

When the number of items to be counted is beyond their number range, children cannot efficiently count. You may see them counting the counter more than once, or skip counters because they do not have a way to organize how they count.

Early Addition and Subtraction

Materials: color tiles or other counters, linking cubes, number cards

Assessment

Seeing Groups

➢ Recognize amounts to 5

Using counters, dot cards, and 5 frames, identify whether or not the child is able to recognize amounts from 1-5 without counting. Instant recognition is obvious, since they will instantly tell you how many they see. When they have to count, there is a pause, and you can often see the eyes moving from object to object. On the recording sheet, write the largest number is instantly recognizable by the student.

➢ Identify and combining parts

Using cards with dot arrangements from 4-10 dots, ask the child how many s/he sees and to tell you how s/he knows. Start with 6 dots. Note whether s/he instantly recognizing, counting all, or seeing groups, such as 4 and 2 more. If s/he is seeing groups, note how the total was calculated, counting on or adding. If s/he counts all, then go to a card with fewer dots. If s/he counts on or adds, the move to a different arrangement with the same number or more dots.

    (3) Independent Level – Instantly recognizes or adds two parts
    (2) Needs Practice – Sees parts and counts on, or inconsistent
    (1) Instructional Level – Counts all
    (0) Frustrational Level – Unable to state total

Going from one number to another

Show student a number card, such as 6. Ask him to build a pile of tiles to match the number you show. Then show a new number, such as 10. Ask him if to make this number they would need to get more or take some away. Ask how many they would need to get or get rid of, then ask them to check and see. With their new pile, show a number such as 7 and ask if to make this number they would need to get more or take some away. Ask how many they would need to get rid of, then ask them to check and see.

   Tells if more or less is needed (score more and less individually)

    (3) Independent Level – Consistently accurate
    (2) Needs Practice   - Accurate with smaller numbers but not with larger
    (1) Instructional Level - Need to build to find out
    (0) Frustrational Level -Unsure

Tells how many more/less (score more and less individually)

     (3) Independent Level - Consistently accurate, just knows by adding/subtracting
     (2) Needs Practice   - Inconsistent, counts on, counts back
     (1) Instructional Level - Need to build to find out, counts all
     (0) Frustrational Level - Unsure

Number Relationships

➤ +1

Provide a train of 4 cubes and ask how many there are. Next to it show a train of 5 cubes, covering the bottom cubes so the student cannot count. Say, "If there are 4 cubes in this train, how many are in this one." Note if students guess or try to count by imagining that they can see the cubes. When they respond, ask how they know.

➤ -1

If the student was accurate and use the relationship of the two trains to determine the amount then show a tower of 6 and ask how many they are. Next to it show a train of 5 cubes (of a different color than the one above). Cover the bottom cubes so the student cannot count. Say, "If there are 6 cubes in this train, how many are in this one." Note if students guess or try to count by imagining that they can see the cubes. When they respond, ask how they know.

➤ +-2, +-3

Continue by providing trains that are 2 more, 2 less, 3 more, 3 less. They may be more accurate with adding than subtracting, so if they miss a subtracting one, stop with that, but continue to do adding relationships.

On the score sheet, place a + on the ones the child is successfully able to do.

Seeing groups

As children are expanding their ability to see quantity, we need to encourage them to find those quantities in a pattern of larger amounts. Prepare a set of dot cards with different patterns and amounts. For example:

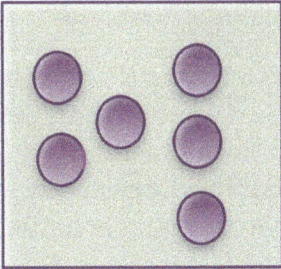

When showing this card to students, we would not necessarily expect them to instantly recognize that there are 6 dots, however, they should easily see 3 and 3, or 5 and 1. Sometimes children say that they saw 2, 2, and 2, but in this case it is most likely that they see the groups, but are counting all to find the total and are not using the groups to find the total amount.

When using dot cards such as these, keep asking students what numbers they see. Model, or have them write, addition sentences that match.

Challenge students by flashing the cards (or patterns on a presentation slide of an interactive white board) for a brief period of time and have the student have to remember the pattern and tell the groups they saw.

Comparing numbers & Number Relationships

In order for children to be able to determine if they need to get more or take some away when going from one quantity to another, they need to know more and less. Innately children know about more with very small amounts, and especially when food is involved. When providing experiences with more and less, we need to be helping the child to develop good comparing language.

Present two piles or trains of counters, or something that s/he wants to have. Give one pile or train for each of you. In the beginning have one pile/train be one more than the other. Ask the child who has more and what they need to do so you both have the same. Have him do it and tell you what he did. Start small if necessary and gradually increase the number size. As he is ready, have the child start to say how many s/he will get before he does it. Connect what s/he does to a number sentence.

When the child is able to add more, start asking how many he should take away so that you two will have the same amount. Connect what he does to a number sentence. Gradually increase the difference between the two amounts.

Show one train, or pile, and a number card. Have the child decide if the number is more or less and how much he will need to get, or get rid of, to even the two amounts. Move toward two numerals and have the child prove by building.

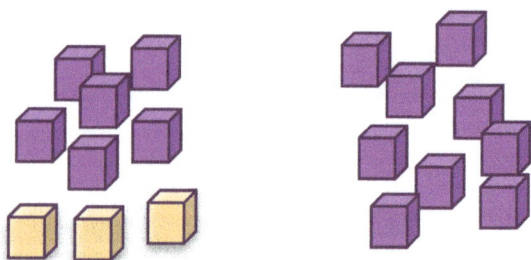

6 + 3 =

Early in the process you want the child to be able to articulate that they know, for example one more than 5 is 6, because 6 is the next counting number. They should start using this rule to know the size of one train by comparing it to another. They should be able to make these generalizable rules for +-1 and +-2. Numbers beyond that will involve counting on or learning the combinations. Work in the area of combinations of numbers to ten will also support the learning of these addition facts.

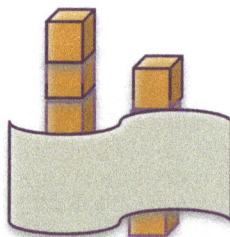

⁺Tens and Ones

Materials: color tiles or other counters

Assessment

Provide a pile of cubes in front of the child and ask him to guess how many there are.

1) Estimate
   - (3) Estimate very close to exact total of tiles
   - (2) Reasonable estimate
   - (1) Estimate way off
   - (0) No estimate

Take his estimate and say, "how many would we have if you counted them by 2's?" (wait for response), "check and see." If they count in groups of 2, when they are done, ask how high they can count by 2's.

2) Counts by 2's – regardless if the child is counting in groups or individual cubes, notice their ability to count by 2's

   - (3) Can count by 2's with ease and accuracy
   - (2) Can count by 2's but skips some numbers or stumbles
   - (1) Can count by 2's to about 10 or 12
   - (0) Unable or inaccurate

Say, "how many would we have if you counted them by 5's?" (wait for response), "check and see." If they count in groups of 5, when they are done, ask how high they can count by 5's.

3) Counts by 5's – regardless if the child is counting in groups or individual cubes, notice their ability to count by 5's

   - (3) Can count by fives to 200
   - (2) Can count by fives to 100
   - (1) Can count by fives somewhat
   - (0) Unable or inaccurate

4) Conservation of number
   - (yes) Child demonstrates conservation of number by stating that they would have the same amount of tiles.
   - (no) Child does not demonstrate conservation of number since he counting each tile by 2's or 5's

"Before when you counted you said there were ____. If so, how many groups of tens would you be able to make?"   Note if the estimate is reasonable and if the number of tens match the estimate.

>   5) Estimate matches number of tens
>   (3) Number of tens matches estimate, and provides number of left overs
>   (2) Number of tens matches estimate – no left overs provided
>   (1) Number of tens does not match estimate
>   (0) Unable to answer

 Ask student to check and see.

>   6) Making tens
>   (3) Makes groups of tens and left-overs, counts them as tens and adds left overs
>   (2) Makes groups of tens and left-overs, counts all by ones or other increments besides tens.
>   (1) Counts all re-evaluates estimate without regard to tens.
>   (0) Counts by ones, but is inaccurate

Ask student to count by tens by rote to 200

>   7) Counting by tens
>   (3) Can count by tens to 200
>   (2) Can count by tens to 100
>   (1) Can count by tens somewhat
>   (0) Unable or inaccurate

With pile of tiles separated into tens and leftovers, ask how many there would be if they had 10, 20 more, if they took 10 away, 20 away.

>   8) Adds / subtracts groups of tens
>   (3) Is able to add tens without counting out more groups
>   (2) Is able to add tens by counting out more groups
>   (1) Counts out more tens, but tells only how many tens (forgets leftovers)
>   (0) Unable or inaccurate

End of Test

If student does well on this test, depending on the age, you may choose to move on to the 2-digit addition and subtraction assessment. That test uses base ten blocks rather than individual cubes, so the increased level of abstractness may cause a shift in accuracy.

Estimating

Estimating is a difficult skill for many students. They are so used to wanting to get an exact answer that they will tend to find the total and round to get an estimate. It is hard for some to let go and just take a guess.

It is important that children get a variety of different experiences to estimate. They need different containers and sized objects with which to estimate. They need to come to the conclusion that if there are bigger objects, there will be less of them that fit in the same size container. This big idea comes back in measurement and fractions.

Weekly Estimation Jars

At least once a week provide 3 estimation jars for students. One jar should be easy, one medium and one challenging. This will allow all students a chance to be successful. You may choose to allow all students estimate the items in all jars, to choose which one they want, or to assign the jar that they estimate.

One helpful strategy is to have three same-size jars, each filled with the same object, but of different amounts. Amounts for two of the jars are labeled, and the third one is not. Students have to use the information they know to guess the third jar.

Materials: color tiles, paper

Estimation of Area

Take square tiles and create an irregular shape (quantity will depend on age level). Trace the shape on paper and cut out. Create several of these shapes. Place one of these shapes on the document camera to project in front of the class. Then place one tile somewhere on the shape so students can see relative size to each tile. Have them make estimates of how many tiles it would take to fill out the entire shape. Place another tile or two and see if the students want to change their predictions. This can be done with even very young children by using smaller figures. This activity is very helpful in developing visual-spatial relationships and estimation.

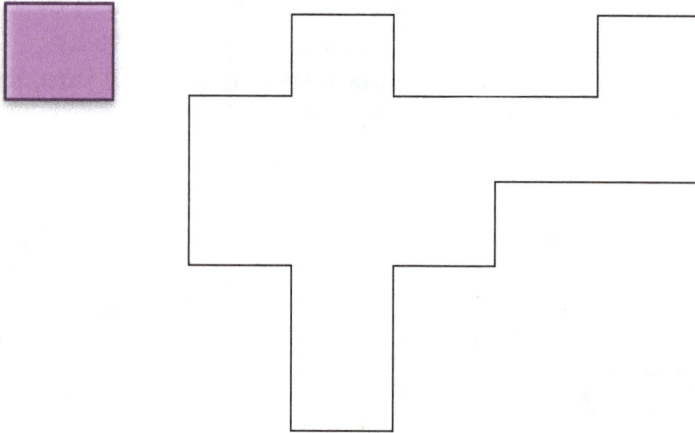

Materials: Large number of cubes to count.

Making Groups of Ten

Dump out a large box of cubes and ask students to count them. Allow them to start counting one-by-one if they wish, but after awhile notice if they start to loose track. They will most likely try strategies to help them count, such as counting by twos. They will still loose count after so many. Ask students if there are ways to group them that will help them organize their thinking so they can keep better track. Some may try fives and others tens. If the child is using fives, let him and eventually when counting back, have him notice that every two groups of five is a ten, so he might want to put two fives closer together to then count them at the same time. If they are counting linking cubes, they can put ten together to make trains. If there are a few hundred cubes, then have them figure out if there is a way to organize their trains so they can keep track. Some students will realize that they can organize their tens trains by putting them into groups of ten .

Connecting the amount of tens of cubes to a number

Students might be able to group tens, but that does not mean that they are making the connection that the number of tens is the first digit in a two-digit number.

<u>Ten Frames</u>
Firstly, students need a lot of practice counting by ten with both objects and numbers. An ideal initial experiences in kindergarten and first grade are with the routine of counting the number of days in school. Since there are 180 days, have posted 18 blank ten frames. Each day when a dot is added to the ten frame, flip the number of tens and ones to match. When a ten frame is filled, off to the side write the ten it represents. This repeated daily exposure will help students make the connection.

Days of School

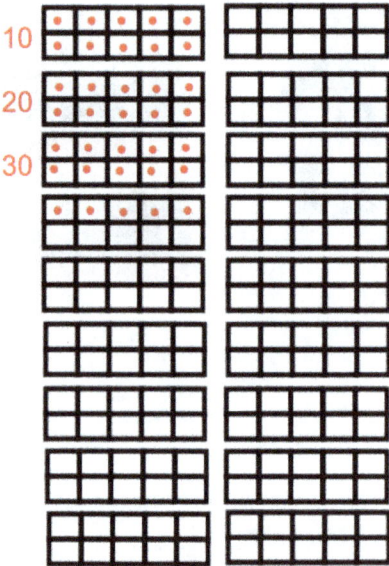

<u>Place Value Mat</u>

When the luxury of time is not something you have, especially when working 1-1 with a student, then simply show them a number and have them build it. You might want to take the cards from a 100 chart and mix them up for the student to pick. If they just count out one-by-one, then encourage them to break it up into groups of ten. On a place value chart, have him put the tens (either as a train or loosely in a cup) and write the number of tens beneath it and the number of ones beneath the left overs. After just a few numbers, the connection should be made. To be sure, pose a problem such as: if you had 63, how many tens and how many left overs. If the child does not know, he needs more time building.

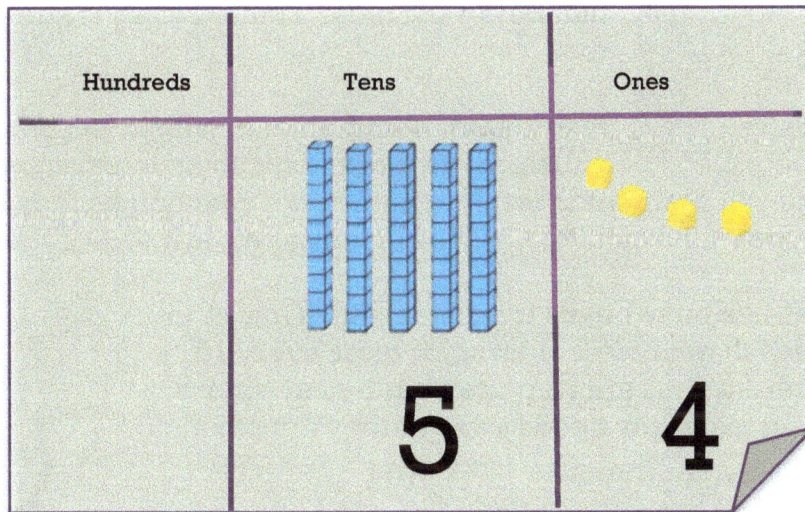

Hundreds	Tens	Ones
	5	4

Skip Counting

Skip counting is something that too many of us take for granted. We think that if children can skip count by rote, they understand what it means. It is common in the classroom for children to even point to a 100 chart when they are skip counting, but that does not mean that they really understand what is happening.

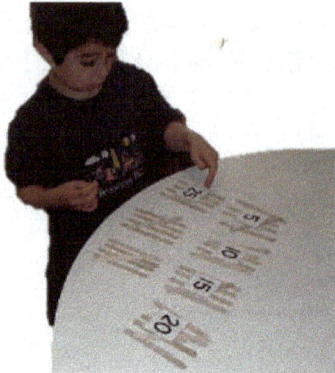

We want children to skip count with meaning. When provided with a variety of objects and asked to count by ten, young children will count each individual object by ten, even if they have been using a 100 chart and only pointing to the tens. They do not realize that they are really skipping nine numbers each time. The 100 chart is still too abstract of a tool.

Start by providing objects such as popsicle sticks or counting cubes. Have the student count out groups of ten. Then have them take the 100 chart cards on them as they are counting so they can see that 10 means ten cubes, 20 means two groups of ten cubes, etc.

They can also use pony beads for this as well. On a string, lace 100 beads of two colors, alternate colors every 10 beads. When students are skip counting by any amount, have them pull that many beads away as they are counting.

Developing Conservation

Conservation of number develops in children when they are about 7-8 years old. If your student still has not seemed to develop conversation, you can provide explicit experiences that can help that along.

Dump a small pile of cubes, or other objects, in front of the child and ask him to count how many he has. Ask him how many he will have if he puts them into groups of twos and count them. Then have him put them into groups of two and ask how many he has now. He may or may not be able to count by twos, but even if he counts by one, he will see that he gets the same amount. Ask how many he thinks he will have if he puts them into groups of five....three...etc. If the student notices that he always has the same amount, ask how he knows. Ask if that would always happen and try with a different number.

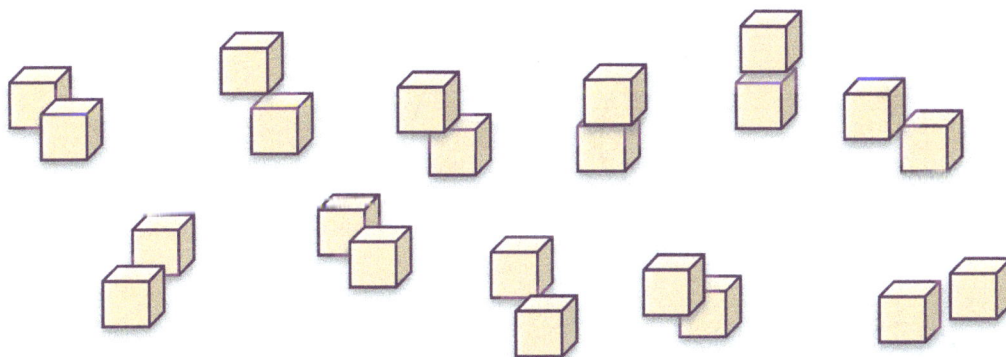

"No matter how you count it there will be the same amount of cubes because you are not adding more or taking any away."

Developing a reason for using 10

Initially, when asked to model a two-digit number, say 24, children are going to create a train or pile of 24 cubes or other objects. We are going to want to give them lots of opportunities to build these two digit numbers, but first we need to give them opportunities to group objects in many different ways. When grouping, the same amount of objects need to be in each group. These activities will also support repeated addition and early multiplication ideas.

Give children lots of things to count, sometimes directed and sometimes independent, students will create groups of a certain number of objects. They should express the number of objects in the amount of groups and left-overs. For example, working with the number 28:

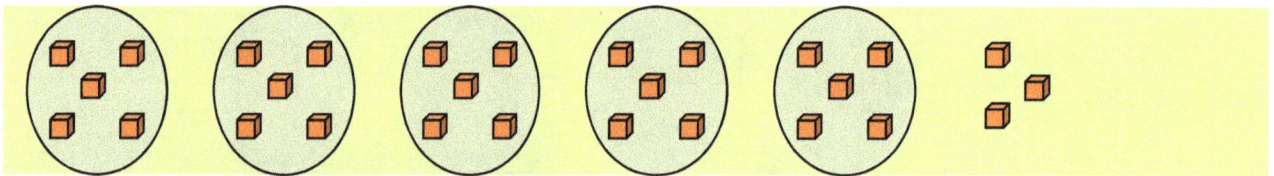

"There are 5 groups of 5 cubes and 3 left-overs."

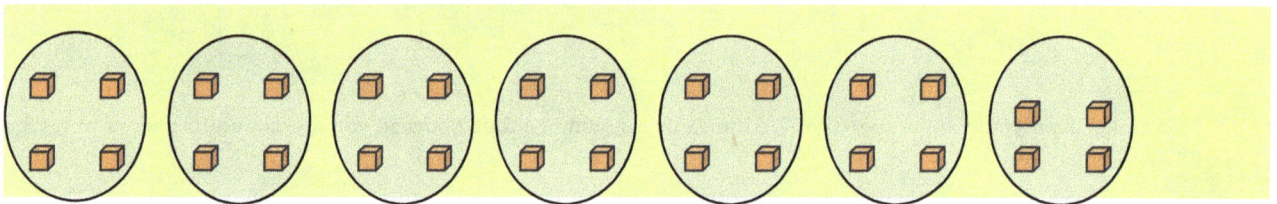

"There are 8 groups of 4 cubes."

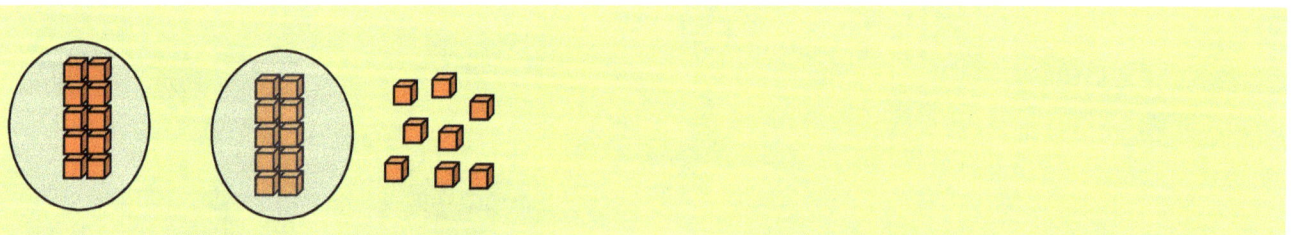

"There are 2 groups of 10 cubes and 8 left over"

If the student were to group 28 objects, as in the above example, then he would fill in his recording sheet as such:

	28	
# in each group	# of groups	# of extras
5	5	3
4	7	0
10	2	8

When the students have done many of these, then through small group and whole group conversations they will begin to discover that when objects are grouped in tens, the number of groups and extras are the same as the way the number is written.

Multi-digit Addition and Subtraction

Materials: Base ten blocks

Assessment

There are several developmental milestones students achieve as they are learning how to add and subtract multi-digit numbers. This assessment determines what the child is able to do and how s/he deals with tens and ones. This will assist the teacher in diagnosing where the breakdown may be occurring to better assist the child.

Use the assessment worksheet and write notes in the Multi-digit Addition and Subtraction section regarding student responses and strategies. Then mark on the score sheet if child is able to successfully perform each milestone.

1) <u>Counts multiples of 10 and leftovers</u>
Present a pile of base ten blocks or unifix™ cubes organized into tens and ones totaling 65. Note how child counts.

    (3) Uses tens and ones
    (2) Counts ones first and adds on tens
    (1) Counts all...not thinking in tens and ones, skip to the computation section.
    (0) Inaccurate...make a note and skip to the computation section.

If the child scored a (0), go back and complete "Counting in Groups" assessment to determine if the child can make groups of tens and left overs and if s/he has developed conservation of number. Assessing the questions in the computation section will let you know if the student has procedures even if there is no understanding of tens and ones.

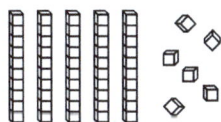

2) <u>Adds multiples of tens without regrouping using concrete/visual model</u>
Using the 65 blocks say, "how many would there be if you added **20** more?"
If 20 is too much, ask, "how many would there be if you added **10** more?"

    (3) Can solve mentally by imagining the tens being added,
    (2) Needs to grab two more tens to see the total,
    (1) Counts 20 ones, either with blocks or on fingers,
    (0) Gives an incorrect answer.

3) **Adds tens and ones without regrouping using concrete/visual model**
Using the same 65 blocks say, "how many would there be if you added **12** more?"
For some reason, with this question children think 12 more tens. Note their
response, then rephrase by saying, "not 12 tens, just 12".

   (3) Can solve mentally by imagining the tens and ones being added,
   (2) Needs to grab a ten stick and two ones to see the total
   (1) Counts 12 ones, either with blocks or on fingers
   (0) Gives an incorrect answer

4) **Adds tens and ones with regrouping using concrete/visual model**
Using the same 65 blocks say, "how many would there be if you added **9** more?"

   (3) Can solve mentally by imagining nine being added
   (2) Grabs a ten stick and covers up one
   (1) Counts out 9 ones, then counts all, or count on from 65
   (0) Gives an incorrect answer

5) **Subtracts multiples of tens without regrouping using concrete/visual model**
Using the 65 blocks say, "how many there would be if you took **20** away?"

   (3) Can solve mentally by imagining the tens being removed,
   (2) Needs to physically remove tens to see the total,
   (1) Counts back 20 starting from the ones, either with blocks or on fingers,
   (0) Gives an incorrect answer.

6) **Subtracts tens and ones without regrouping using concrete/visual model**
Using the same 65 blocks say, "how many would there be if you took **15** away?"

   (3) Can solve mentally by imagining the tens and ones being removed,
   (2) Needs to remove a ten stick and 5 ones to see the total
   (1) Counts back 15 starting with ones, either with blocks or on fingers
   (0) Gives an incorrect answer

7) **Adds tens and ones with regrouping using concrete/visual model**
Using the same 65 blocks say, "how many would there be if you took **9** away?"

   (3) Can solve mentally by imagining removing a ten and replacing a one
   (2) Covers nine from a ten stick and replaces one
   (1) Counts back 9 starting from ones with blocks or on fingers
   (0) Gives an incorrect answer

Computation

Present the following problems to the student written horizontally. Ask him/her to solve in any way that they choose. Note if they rewrite the problem vertically. If they solve incorrectly horizontally (for example 28 + 24 = 412), try writing it vertically to see if it changes the way they respond. Ask student, "can you show the way you solved it with numbers by using the blocks?" You are looking to see if they can connect their numerical strategy to a concrete/visual model.

$$14 + 12$$
$$29 - 14$$
$$28 + 24$$
$$34 - 16$$
$$281 + 241$$
$$376 - 198$$

(3) Solves accurately both horizontally and vertically, and is able to connect to model

(2) Solves accurately both horizontally and vertically, but is unable to connect to model

(1) Solves accurately vertically, but is unable to connect to model

(0) Inaccurate

Mental computation

Present the following problems and ask the student to solve them in his head.

$$1000 - 998$$
$$99 + 16$$
$$15 + \_\_\_\_ = 200$$

(3) Able to solve all 3 problems with reason and use of relationships

(2) Able to solve 1 or 2 problems with reason and use of relationships

(1) Solves using standard algorithms

(0) Inaccurate

Creating word problems for addition and for subtraction

Ask, "Create your own story which would use addition / subtraction"

(3) Able to create a word problem that matches either addition or subtraction with an appropriate question

(2) Able to create an appropriate situation, but doesn't include question

(0) Unable to create a word problem that matches either addition or subtraction

Building foundations for regrouping

It is very important that we remember when teaching the 4 operations, to start from the concrete and gradually move toward the symbolic by making explicit connections between the concrete, pictorial, and symbolic notations. When moving away from concrete to more abstract representations, be careful that understanding is not lost. If children start doing random things to numbers, then you need to immediately go back to a more concrete representation until connections are stronger.

Consider the Learning Cycle Below. First we must develop understanding by constructing the mathematics with a concrete and/or pictorial model to make sense of the problem. Then we must connect it to a numerical representation so that the procedures make sense. It is through these connections that students can use their right hemisphere to help build the areas in the left associated with computation and symbolic notation.

The sequence for developing understanding around addition and subtraction with regrouping are on the following pages. Starting from the beginning will ensure that there are no gaps. The following pages will move students from learning how to build, draw, to recording addition and subtraction problems numerically.

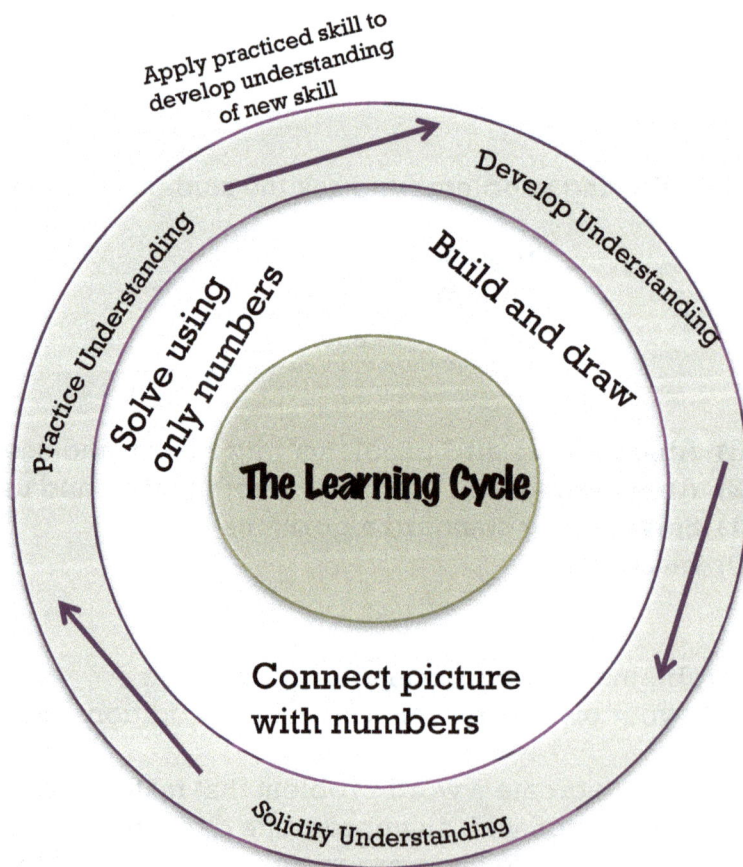

The Learning Cycle, adapted from the work of Bahr, Sterling, and Hilton at Brigham Young University.

Representing a Number with Base Ten Blocks
(or Unifix™ cubes if base ten blocks are still too abstract for the child)

"How can you represent 24 using base ten blocks?"

"How can you draw a picture of that?" Use simple lines for tens and dots for ones.

Play Race to 50 (or race to 100)

Use a place value mat, unifix™ cubes or base ten blocks, and a die

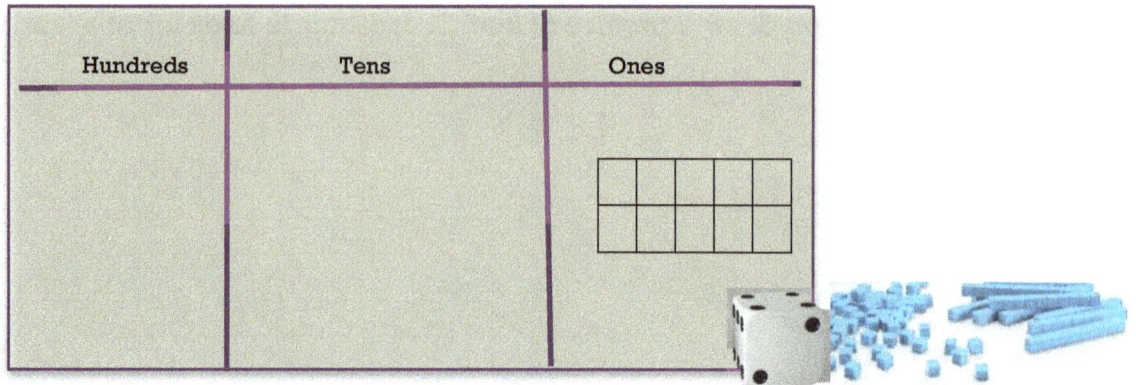

Hundreds	Tens	Ones

The point of this game at this time is for the child to become comfortable with trading every time they have at least ten. Have the student roll a die and place that many ones in the ones section on the place value mat. Put one cube in each of the spaces on the ten frame. This will help remind students to trade when the ten frame is full. Keep rolling the die and adding cubes. When the frame is full, the student is to trade for a ten stick and move it to the tens. Repeat until the desired amount (50 or 100) has been reached.

If the ten stick is too abstract, put the ten cubes into a small cup and place in the tens space. Or use unifix™ cubes where students can manually connect the cubes into a ten train.

Misconceptions: When having a number such as 12, student may trade all 12 for the ten stick. It is very common that they do not trade 10 for 10.

Materials: base ten blocks,

Adding tens and ones to a number
No regrouping required

→ Present 3 ten sticks

   "imagine you have 10 more...20 more...30 more..."
   Do this a variety of times with a different amount of tens. If the child cannot visualize, allow him to get the additional tens.

|||

→ Present 3 ten sticks and two ones

   "Imagine you have 10 more...20 more...30 more..."
   Do this a variety of times with a different amount of tens and ones. If the child cannot visualize, allow him to get the additional tens...eventually hold up the additional tens away from the original amount, then fade out the second quantity and can just show numeral if necessary.

||| °°

→ Present 3 ten sticks and two ones

   "Imagine you have 11 more...12 more..."
   Do this a variety of times with a different amount of tens and ones. If the child cannot visualize, allow him to get the additional tens...eventually hold up the additional tens away from the original amount, then fade out the second quantity and can just show numeral if necessary.

||| °°

→ Have the child add 10, 11, 12, etc. to a specified number on the 100 chart to see patterns and develop an additional mental structure for adding these numbers. Students will most likely count up on the 100 chart...let them. They will need to count until they discover the pattern for themselves. Help them make it obvious by marking the starting and ending spot with a counter or dry erase marker.

1	2	3	4	5	6	7	8	9	10
11	12	13	14	15	16	17	18	19	20
21	22	23	24	25	26	27	28	29	30
31	32	33	34	35	36	37	38	39	40
41	42	43	44	45	46	47	48	49	50
51	52	53	54	55	56	57	58	59	60
61	62	63	64	65	66	67	68	69	70
71	72	73	74	75	76	77	78	79	80
81	82	83	84	85	86	87	88	89	90
91	92	93	94	95	96	97	98	99	100

Addition without regrouping
Build, draw, and connect to numbers

→Connect blocks to picture

Pose problems such as 33 + 24 and have student build both numbers with the base ten blocks, combine and then draw a picture of what he built and how he combined the blocks.

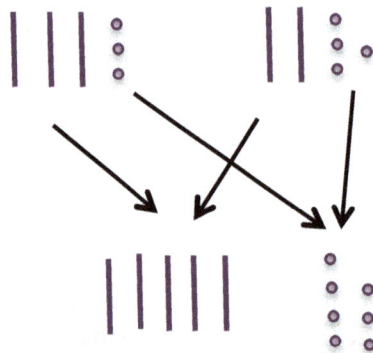

→ Connect blocks and pictures to numbers

Pose similar problems, but now add in the recording of numbers.

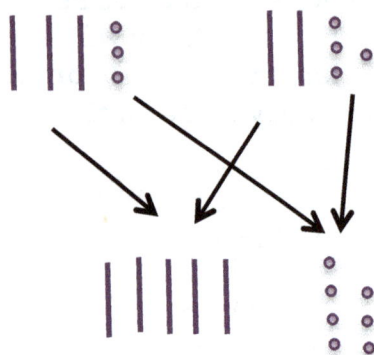

$$33 + 24$$

$$50 + 7 = 57$$

→ Solve 2 digit addition problems mentally by thinking of tens and ones.
Pose problems horizontally and encourage students to solve mentally in the same manner as above.

→ Build, draw and solve 3 digit addition problems mentally by thinking of tens and ones.

Addition with regrouping
Build, draw, and connect to numbers

Materials: base ten blocks,

→Building addition with blocks

Pose problems such as 38 + 24 and have students build with blocks. Most students will lay out both numbers, then put all the tens together and all the ones. They will even most likely count the tens and count up all the ones to get 62. If they do, ask, "how many ones do you have now?" This should be enough for them to trade a ten, but if not, ask what they do when they have more than ten ones. Pose several problems to ensure ease in building before moving on to pictures. Use the place value mat if student needs more support

→Connect model to picture

Pose problems such as 38 + 24 and have students build both numbers with the base ten blocks, combine and then draw a picture of what he built and how he combined the blocks. Do several examples before moving to the next step.

Example:

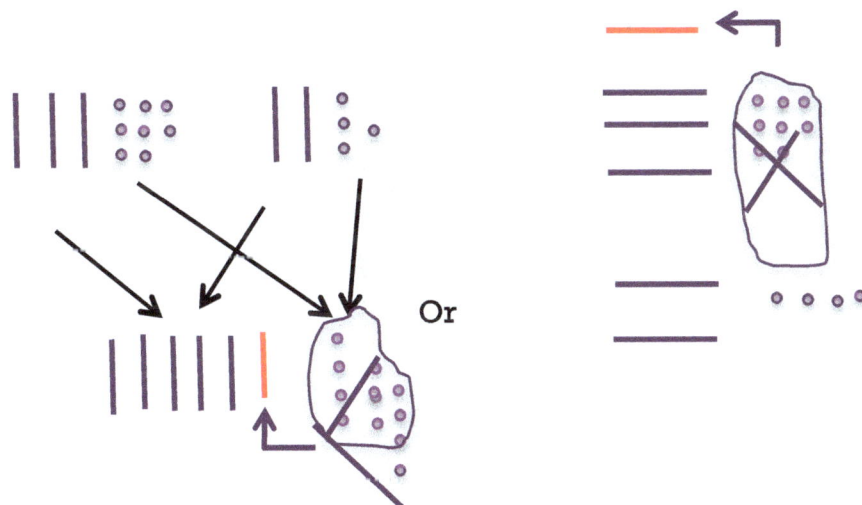

Or

→ Connect model and pictures to numbers

Pose similar problems, but now add in the recording of numbers. There are a couple of ways to record the number, depending on how the blocks are organized. If he is still not trading the ones for an extra ten, that can still be recorded numerically:

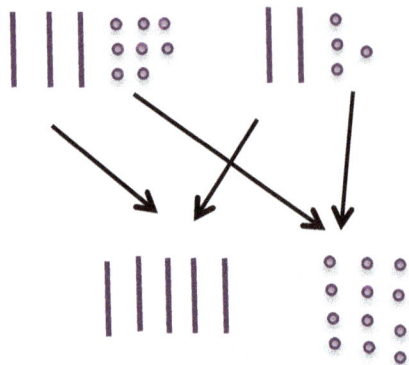

$$38 + 24$$
$$50 + 12 = 62$$

The above representation is good when trying to move towards the partial products algorithm that preserves place value. If the purpose of the modeling is to move towards the standard algorithm, the recordings might look like one of the following (from most concrete to most abstract):

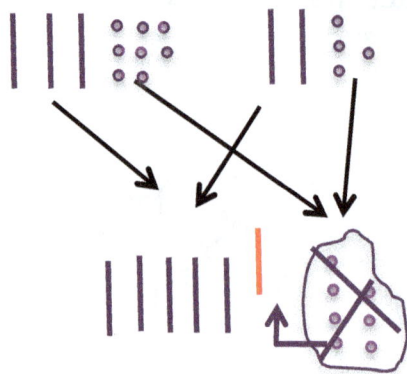

$$38 + 24$$
$$50 + 12$$
$$50 + 10 + 2$$

$$38 + 24$$
$$50 + 12$$
$$60 + 2$$

OR

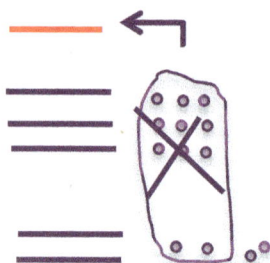

$$\begin{array}{r} {}^{10} \\ 30 + 8 \\ +20 + 4 \\ \hline 60 \; + 2 \end{array}$$

$$\begin{array}{r} {}^{1} \\ 38 \\ +24 \\ \hline 62 \end{array}$$

When children are working towards the standard algorithm, it is important to realize that the notation is a shortcut and students need to make sense of the long cut first before the short cut really makes sense.

→ Solve 2 digit addition problems mentally by thinking of tens and ones.

Pose problems horizontally and encourage students to solve mentally by adding tens then ones.

→ Build, draw and solve 3 digit addition problems mentally by thinking of tens and ones.

Follow the sequence used for 2-digit addition.

Materials: base ten blocks, dice, place value chart

Play Race to zero

Use a place value mat, unifix™ cubes or base ten blocks, and a die

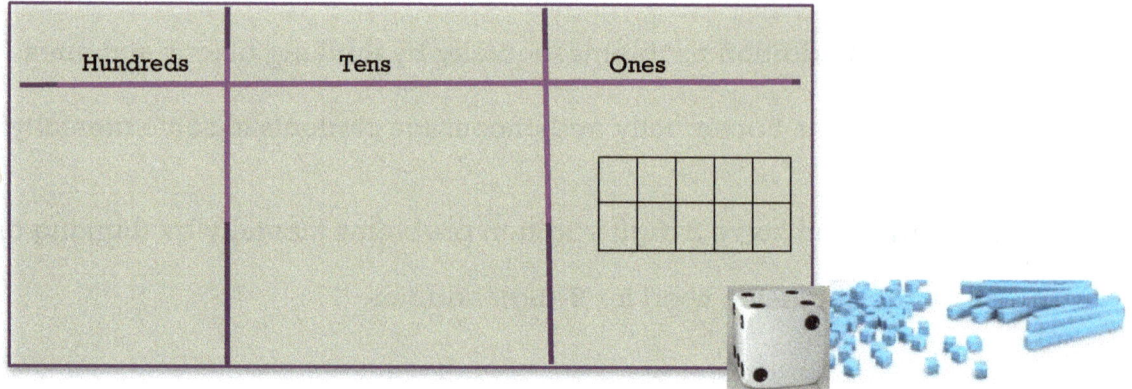

Hundreds	Tens	Ones

This is similar to playing race to 50, but in reverse. Start with either 5 tens or one hundred block. Student rolls the die and has to remove that amount. The first time they play, they will have no idea how to remove, lets say 3, from 50. They may try to take away 3 tens or just cover up three ones from a ten. Remind them that they can only take away from ones, so as how they might be able to do that. By then, they should realize that they need to trade. If the child is making too many errors with trading, you may decide to first use unifix™ cubes to break apart rather than trading.

Play with a ten die and ones die starting with at least 100.

Misconceptions: If 3 needs to be taken from 50, students may trade a ten and get 3 ones. If there are 4 and 2 ones, they may trade a ten and get one more cube to make the 3 they need to trade for. As before, they may have a hard time trading ten for ten.

Subtracting tens and ones from a number
No regrouping required

→ Present 4 ten sticks

"imagine you take 10 away...20 away...30 away..."
Do this a variety of times with a different amount of tens. If the child cannot visualize, allow him to remove the tens.

| | | | |

→ Present 4 ten sticks and two ones

"Imagine you take 10 away...20 away...30 more..."
Do this a variety of times with a different amount of tens and ones. If the child cannot visualize, allow him to remove the tens.

| | | | | °°

→ Present 4 ten sticks and two ones

"Imagine you take 11 away...12 away..."
Do this a variety of times with a different amount of tens and ones. If the child cannot visualize, allow him to remove the tens and ones.

| | | | | °°

→ Have the child subtract 10, 11, 12, etc. on the 100 chart to see patterns and develop an additional mental structure for adding these numbers. Students will most likely count up on the 100 chart...let them. They will need to count until they discover the pattern for themselves. Help them make it obvious by marking the starting and ending spot with a counter or dry erase marker.

1	2	3	4	5	6	7	8	9	10
11	12	13	14	15	16	17	18	19	20
21	22	23	24	25	26	27	28	29	30
31	32	33	34	35	36	37	38	39	40
41	42	43	44	45	46	47	48	49	50
51	52	53	54	55	56	57	58	59	60
61	62	63	64	65	66	67	68	69	70
71	72	73	74	75	76	77	78	79	80
81	82	83	84	85	86	87	88	89	90
91	92	93	94	95	96	97	98	99	100

Subtraction without regrouping
Build, draw, and connect to numbers

→Connect model to picture

Pose problems such as 68 - 24 and see what students will do. Most will try to build both numbers with the base ten blocks then remove 24, but they are confused when their answer is 68. Remind them what they did when playing race to zero, if necessary. Have the child draw a picture of what he did with the blocks. Pose several problems until recording with pictures is comfortable.

→ Connect model and pictures to numbers

Pose similar problems, but now add the recording of numbers. They will most likely remove the tens first and then the ones. Record in numbers the same way they model.

for example:

$$68 - 24$$
$$60 - 20 = 40$$
$$8 - 4 = 4$$
$$40 + 4 = 44$$

Or

```
   68              68            60      8
 - 24            - 24          - 20      4
 ─────          ─────          ────    ───
   40               4            40      4
 +  4            + 40
 ─────          ─────
   44              44
```

When students are comfortable, tell them that for organizational purposes, we are going to record what is done to the ones first. That way it will mimic the standard algorithm.

→ Solve 2 digit subtraction problems mentally by thinking of tens and ones.
Pose problems horizontally and encourage students to solve mentally in the same manner as above.

→ Build, draw and solve 3 digit addition problems mentally by thinking of tens and ones.

Materials: base ten blocks,

Subtraction with regrouping
Build, draw, and connect to numbers

→ Connect model to picture

It is most helpful to use the place value mat, however if that much scaffolding is not necessary, they can build on the table and draw in their math journals. When using the mat, put in a plastic sleeve and use a dry erase marker. Draw on the mat exactly what they build and do with the blocks. Any time they trade or remove blocks, it should be recorded that by crossing out, not erasing.

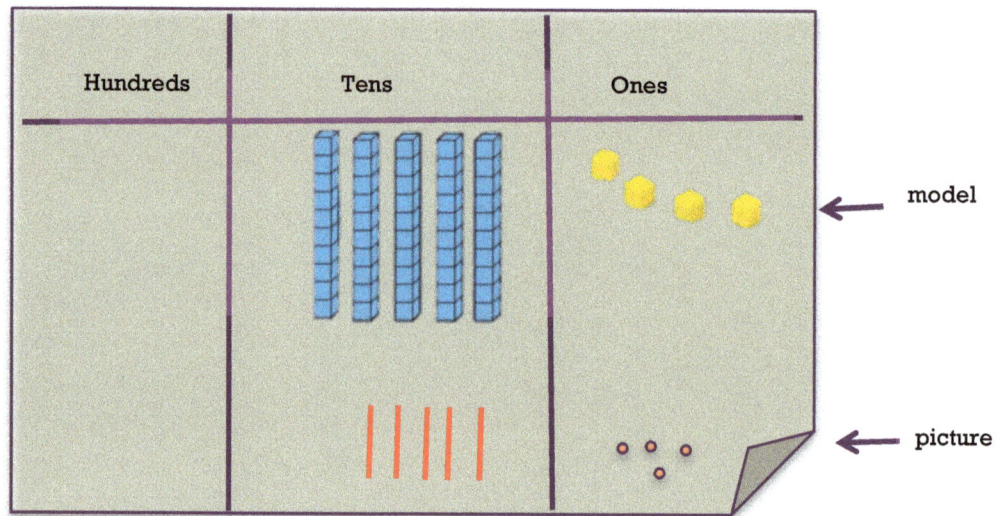

Example: 54 – 26

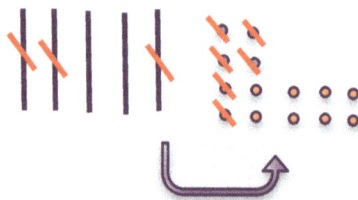

Student Explanation: I removed twenty, then I traded one ten for ten ones. Finally, I removed six ones.

→Connect Model and picture to numbers

Pose similar problems, but now add in the recording of numbers. For sake of organization and to most closely mimic the standard algorithm, say that we are going to start with the ones. He can try recording first from the tens, but he will find that erasing will be necessary after the trade is made. The trick will be to get the student to notice off the bat if trading will be necessary. If so, they should trade with their blocks, record it in the picture, and then write it in the numbers. At first, it will most likely be necessary for the teacher to record in numbers what the student does with the model/picture. Make sure you/he is recording exactly what he is building/drawing after each step.

A)

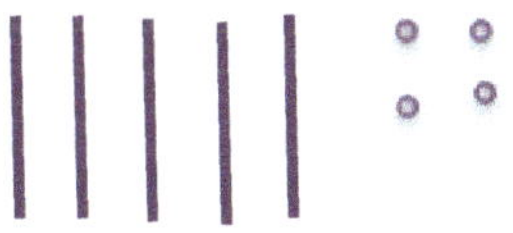

$$
\begin{array}{r}
54 \\
-\ 26 \\
\end{array}
$$

B)

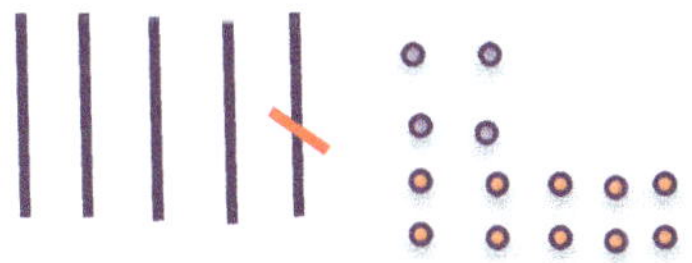

40 14

$$
\begin{array}{r}
\cancel{54} \\
-\ 26 \\
\end{array}
$$

Renaming 54 to 40 and 14...still totaling 54.

C)

40 14

$$
\begin{array}{r}
54 \\
-\ 26 \\
\hline
8 \\
\end{array}
$$

Taking away the ones.

D)

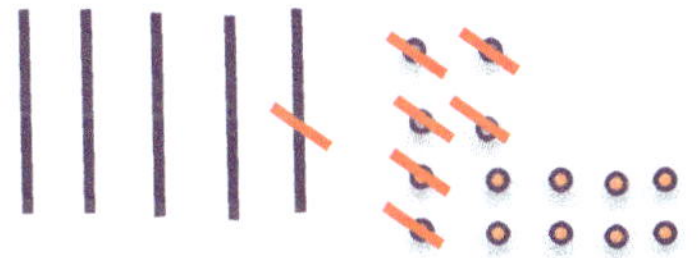

40 14

$$
\begin{array}{r}
54 \\
-\ 26 \\
\hline
28 \\
\end{array}
$$

Taking away the tens.

→ Build, draw and solve 3 digit subtraction problems.
Follow the sequence used for 2-digit addition.

⁺Rounding Numbers / Even & Odd

Materials: Number cards

Assessment

Rounding

Have a variety of number sizes available: 2 digit, 3 digit, 4 digit, and numbers that go to the tenths, hundredths, and thousandths.

When presenting numbers, do two of each, one that will round up and one that will round down. Present randomly to prevent student catching onto a pattern of how numbers are presented.

First present a two-digit number, such as 28. Ask what the number would be rounded to the nearest ten. Present another number, such as 42 and ask the same question. When presenting numbers with a variety of places, ask the places in random, such as rounded to the nearest thousand, ten, then hundred.

→ 2 digit number: "what number would it be when rounded to the nearest ten?"
→ 3 digit number: "what number would it be when rounded to the nearest ten, nearest hundred?"
→ 4 digit number: "What number would it be when rounded to the nearest ten, hundred, thousand?"
→ decimal: "what number would it be when rounded to the nearest whole number, tenth, hundredth, thousandth?

Scoring
+ Student is able to correctly round to requested place
- Student is unable to correctly round to the requested place

Odd/even

Present the number 3 and ask the student whether it is an odd or even number. If he is correct, present the number 12 and then 90. Ask the student what odd or even means and to prove it somehow.

(3) Correctly states 3, 12, and 90 are even or odd, plus can explain why,
(2) correctly states 2 and 12 are even or odd, plus can explain why
(1) Inconsistently states a number is even/odd...has some idea of why
(0) Gives an incorrect answer

Rounding numbers

Materials: pen and paper, interactive white board, white electrical tape or masking tape, linking cubes

Students often struggle with rounding, usually as the numbers get larger because they are not sure what to attend to.

When initially teaching rounding to the nearest ten with two digit numbers, children do not usually understand what rounding really means. Rounding should first be taught within the context of estimating, so there is some reason behind what they are doing. Using a rounding hill is helpful in seeing on a number line which ten, for example, is closest.

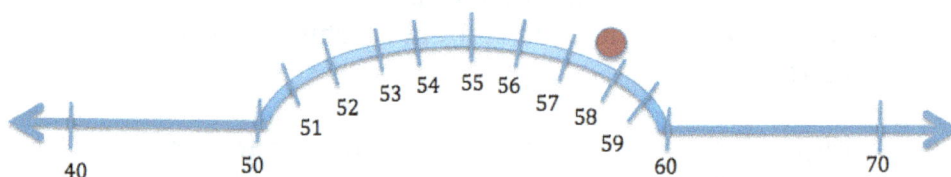

A common error students make is to identify 36 as between 20 and 40. They see the 30 in their target number and go ahead and back a 10. Before even worrying about rounding, some work needs to be done just to identify the two end numbers.

Rounding can even been done with base ten blocks. When building 28, for example, students can easily see that only two more cubes would complete 30 vs. taking 8 away to make 20.

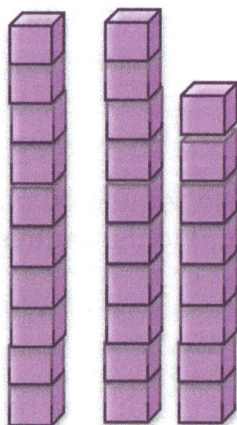

I have also had success with students who have problems with retention, to put a piece of electrical tape across the table and mark hundreds across the tape. In between I put a tic mark representing the 50 and when presenting a number such as 438, I will have the student find where it should fall on the number line, then he can more easily figure out to which ten and/or hundred the number is closest.

Odd/Even

Some students may know that even numbers end in 2, 4, 6, 8, or 0, but not really understand what the word *even* really means. The conjecture that we want students to make regarding odd and even numbers is:

"Even numbers can be broken into two equal groups with no leftovers. Odd numbers always have one leftover when broken into two equal groups."

The easiest way to work with odd and even numbers is to build them with linking cubes. Taking a pile of cubes that match a number presented, such as 8, students will split them into two groups and record what they notice. They should also drawing it on grid paper and cutting it out:

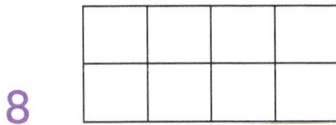

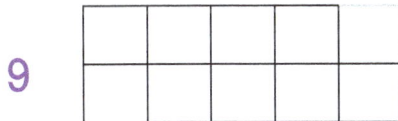

8

9

Although students might have a sense of odd and even with small numbers, when presented a larger number, the misconceptions can show. Take 90. Although this is an even number, the numeral "9" stands out and students will say that it is odd. When asked to prove it, they may grab 9 ten sticks and sort them into two piles and indeed end up with one extra. They need to be asked if the left over can be further broken down. This usually helps students to realize that they can trade for ones.

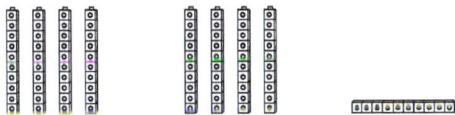

Experiences with larger numbers will help students develop a deeper understanding of odd and even numbers.

Using fingers to start to learn about odd/even numbers

Children who have difficulty with retention and remembering which numbers are odd/even, using their fingers can be helpful. If they learn that they need to look at the ones place, then they only have to build up to 10 digits.

Have the child put up one finger on alternating hands to count up to a number. If all fingers can be paired up, then the number is even.

For example, 6 is even because three fingers on one hand match 3 on the other.

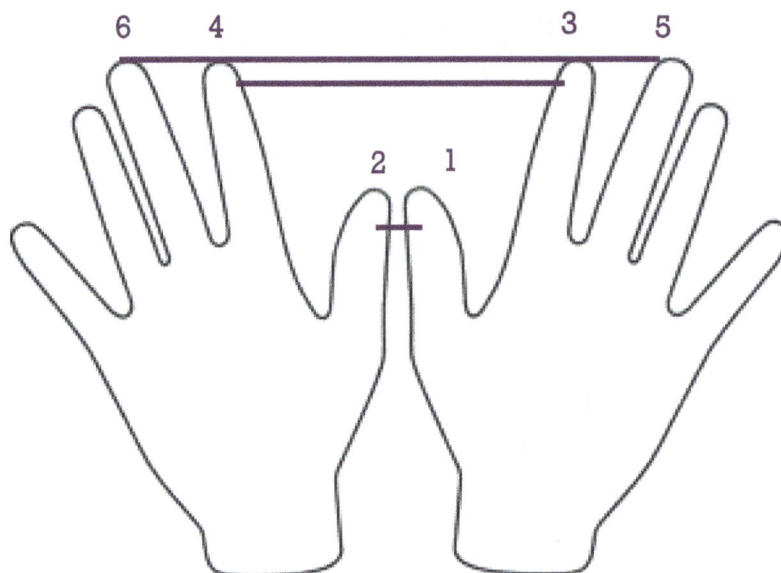

Addition & Multiplication of Odd and Even Numbers

When trying to deepen understanding of odd and even numbers, students can explore what happens when two odd or even numbers are added together. When using manipulatives or pictures, children can easily see that two odd numbers added together make an even number because each odd number has an "extra" and together they make an even number. Two evens also make an even because they did not and still do not have any leftovers. Only an even and an odd are still odd because that extra one does not have anything to partner with.

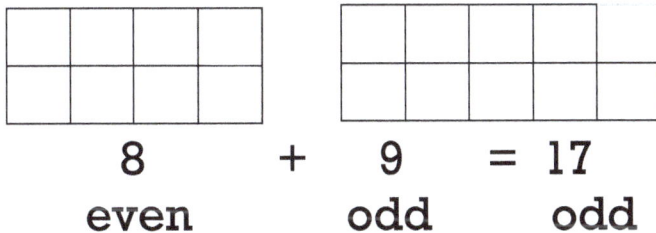

$$8 + 9 = 17$$
even   odd   odd

They can also explore what happens when two even, two odd, or an even and an odd number are multiplied together?

This idea can to be explored in about fourth grade, or once multiplication is understood. Two evens have an even product because there are never any leftovers. An even and an odd are even because no matter how many groups of an even you have, the number is still even. An odd and an odd are the only combination that can produce an odd number because if there were an even group of odd numbers the leftovers would be paired. In an odd number of groups, there is still one leftover.

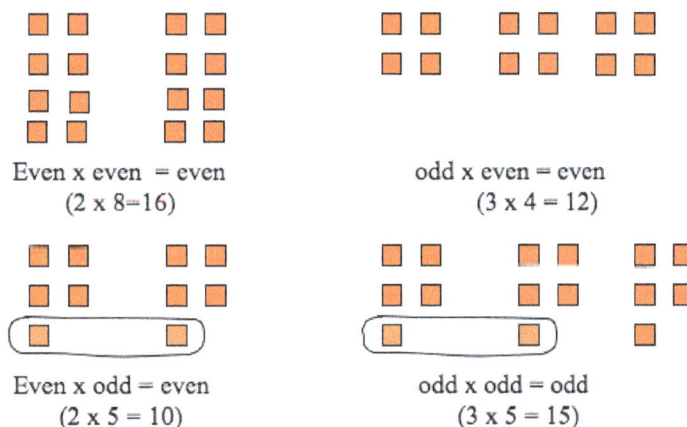

Even x even = even
(2 x 8 = 16)

odd x even = even
(3 x 4 = 12)

Even x odd = even
(2 x 5 = 10)

odd x odd = odd
(3 x 5 = 15)

91

⁺Multiplication and Division

Materials: array cards, pencil & paper, base ten blocks,

Assessment

There are several developmental milestones students achieve as they are learning how to multiply and divide. This assessment begins with conceptual understanding of multiplication and builds towards the ability to solve multi-digit problems.

Use the multiplication and division section on the worksheet and write notes regarding student responses and strategies. Then mark on the score sheet if child is able to successfully perform each milestone.

1)   <u>Connecting multiplication and division to arrays and groups</u>
  Present the different array/group cards and ask:
      "How many squares are there?"
      "How did you count them?"
      "What number sentences can be written about this picture?"
      (see if they write addition, multiplication, or division sentences)

  If the student only wrote addition sentences, ask if a multiplication sentence could be written to match the picture. If s/he only writes one multiplication sentence, ask if another sentence could be used to describe the picture.

  If the student is able to accurately write at least one multiplication sentence to describe the array, then ask if a division sentence could be written about the picture as well.

    (multiplication section)
        (3) Identifies 2 multiplication sentences that match array / groups
        (2) Identifies only 1 multiplication that matches array / groups
        (1) Identifies repeated addition sentence to match
        (0) Number sentences do not match array or group, even if total is same, or unable to answer.

    (division section)
        (3) Identifies division sentences that match array / groups
        (2) Identifies only 1 division sentence that matches array / groups
        (0) Number sentences do not match array, even if total is same, or unable to answer.

2) <u>The language of multiplication</u>
Ask the following questions (located also on worksheet)

> "If I had four fives, how many would that be altogether?"
> "If I had three fours, how many would that be altogether?"
> "If I had six twos, how many would that be altogether?"

(3) Easily and accurately solves the problems
(2) Might need prompt, such as saying, "if I had 4 dice and they all had 5's on them," otherwise accurate
(1) Inconsistently accurate and may need prompt
(0) Gives an incorrect answer.

3) <u>Skip counting</u>
Ask student to skip count by a variety of different numbers: 10, 5, 2, 3, 4, 6, 9. If they know 6, you can choose to test 7 and 8. Mark the ones they are able to do with an X

4) <u>Calculations</u>
Provide the following multiplication problems. Allow students to solve in any way they need. Stop when student is inaccurate or is solving by drawing tallies, etc.

6 x 8	48 ÷ 6
4 x 13	56 ÷ 3
15 x 10	108 ÷ 9
60 x 40	180 ÷ 10
14 x 22	180 ÷ 12

(for each category on score sheet)
(3) Can solve accurately and efficiently,
(2) Uses strategies such as repeated addition to solve, mostly accurate
(1) Uses strategies such as making tally marks or building groups
(0) Gives an incorrect answer

5) <u>Creating word problems for multiplication and division</u>
Ask student to create a word problem for 3 x 4. If successful, ask to create one for 15 ÷ 3.

(grade multiplication separately from division)
(3) Able to create an appropriate situation to match the given problem.
(1) Attempts to create a situation, but it is lacking, (i.e., I had 3 apples, how many would there be if I multiplied 3 x 4?)
(0) Unable to come up with any kind of situation

6) <u>Order of Operations</u>

Present the following problems (located on the task cards), one at a time, stopping when the student makes an error that indicates that s/he does not understand that level of order of operations (errors in computation do not count).

$$4 + 6 \times 2$$

$$3 \times (4 + 7)$$

$$2 + 4 \times 3 + (3^2 - 2)$$

(3) Solves for all three levels,
(2) Solves for the first and second level
(1) Solves the first level only
(0) Gives an incorrect answer

Assessment Task Cards

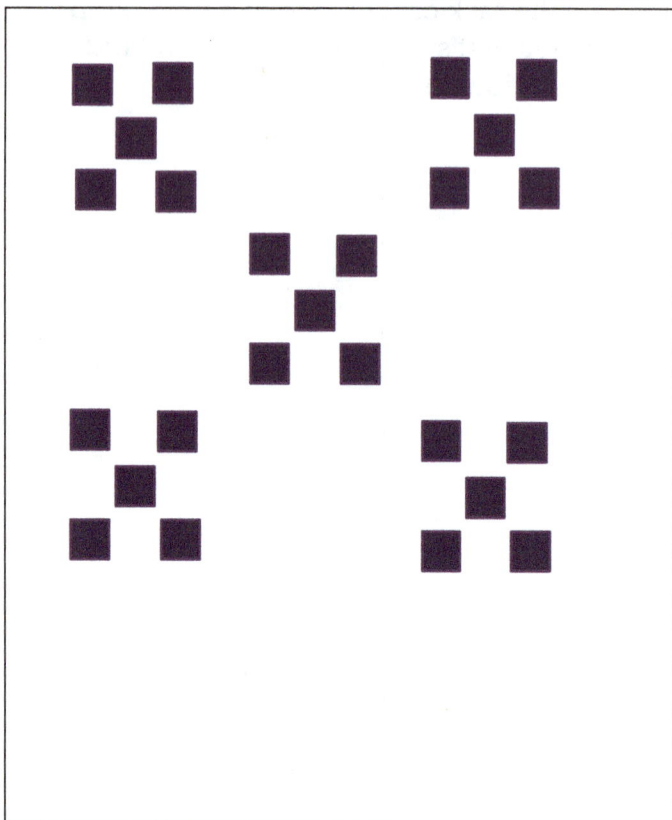

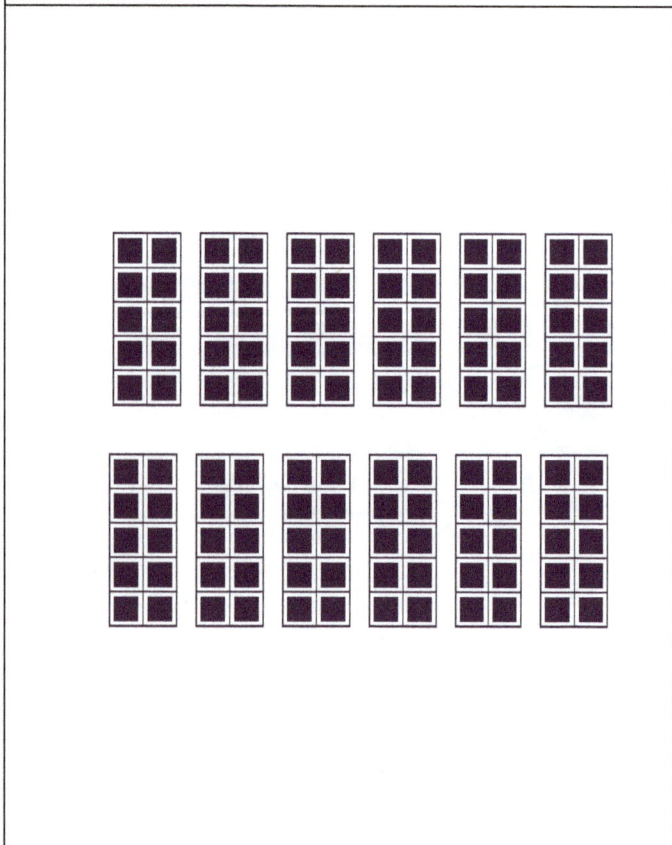

$$4 + 6 \times 2 =$$

$$3 \times (4 + 7) =$$

$$2 \times 3 + (3^2 - 2) =$$

Building foundations for multiplication and division

It is very important that we remember when teaching the 4 operations, to start from the concrete and gradually move toward the abstract by making explicit connections between the concrete, pictorial, and symbolic notations. When moving away from concrete to more abstract representations, be careful that understanding is not lost. If children start doing random things to numbers, then you need to immediately go back to a more concrete representation until connections are stronger.

Consider the Learning Cycle Below. First we must develop understanding by constructing the mathematics with a concrete and/or pictorial model to make sense of the problem. Then we must connect it to a numerical representation so that the procedures make sense.

The sequence for developing understanding around multiplication and division are on the following pages. Starting from the beginning will ensure that there are no gaps. The following pages will move students from learning how to build, draw, to recording multiplication and division problems numerically.

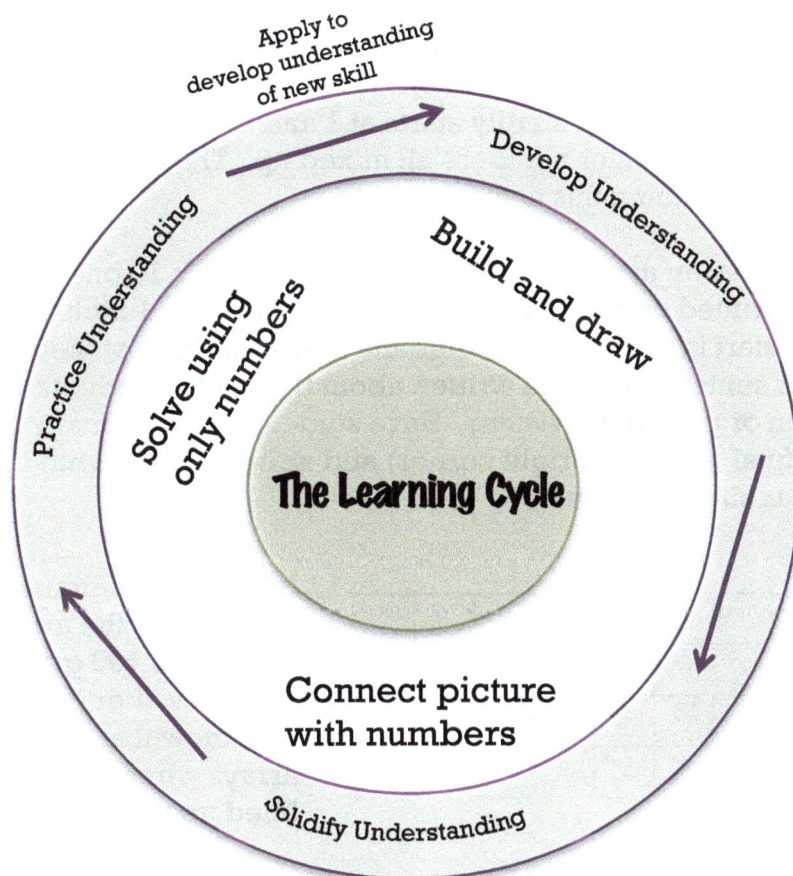

Apply to develop understanding of new skill

Develop Understanding

Practice Understanding

Solve using only numbers

Build and draw

The Learning Cycle

Connect picture with numbers

Solidify Understanding

The Learning Cycle, adapted from the work of Bahr, Sterling, and Hilton at Brigham Young University.

Materials: 1 inch tiles, 1 inch grid paper, markers

Learning the multiplication tables

Any teacher will agree that students who have taken the time to commit their basic facts to memory are more successful in math. Some reasons are that they are less prone to errors, they are able to see connections between operations more easily, and it takes less time to complete tasks.

Teach students how to study their facts. Some children simply do not understand the concept of studying something to the point of memorization. They do not understand that simply looking at a problem does not mean they will learn it. The need to learn strategies such as choosing only a few problems, trying to solve it without seeing the answer and checking to see if they are correct.

Being able to skip count is a key component to memorizing the multiplication tables. If a child cannot skip count, until they memorize the facts, the only solution strategy they have is to count all or draw all the tally marks. A blackline master with skip counting strips is located at the end of this section.

Building the multiplication chart

Many students do not understand how to use a multiplication chart. To them it is a chart with a bunch of random numbers, with no rhyme or reason to its organization. They have been use to a 100 chart that simply starts at 1 and counts in order to 100, but this one has large numbers and smaller numbers all mixed up. They will gain a deeper sense of how to use the chart by doing this activity.

Ask students what an array is. Provide them with a 10 x 10 one-inch grid (or 8 x 10 landscape if printed on regular printer paper) and color 1 inch tiles. They should be instructed to start in the top left corner and make an array of their choosing. Ask them what number sentences can be written about this array (students may provide addition, multiplication or possibly division). Have students tell how many are in total and then remove the final tile (lower right corner) and write that total with a marker on that square. An example is shown below:

3 x 5

Inform the students that they will be building all the arrays that they possibly can on the 100 grid and recording the total tiles after completing each one. They will know that they have completed all the arrays when all of the squares have been filled up with numbers.

What can happen is that students try to fill out the grid like a hundred chart without making all the arrays, but you need to encourage them, at this point, to keep building the arrays. They will soon realize that it will not work if they just make a 100 chart.

After completing about a third to a half of the arrays, they will generally begin to see that they are skip counting, perhaps with the tens. If they start filling in the numbers by skip counting, ask them questions, such as why they are choosing to put that number, what patterns they are noticing, etc.

As a whole class, students should discuss what they notice. Many will notice some patterns, such as the tens. Ask them why there are so many of the same number, such as the 9 or 12. They might not know how to respond, so build one of those numbers and build another. You should guide them to notice that there are many ways to make the same number, and record them on the board (for example: 2 x 6, 3 x 4, 6 x 2, 4 x 3). Ask them what they notice. They should notice that there is a 2 and a 6 in two of the problems, etc. Have students build each one on the chart and allow them to see that the arrays are simply rotated, from one to the other. Invite students to

x	1	2	3	4	5	6	7	8	9	10	11	12
1	1	2	3	4	5	6	7	8	9	10	11	12
2	2	4	6	8	10	12	14	16	18	20	22	24
3	3	6	9	12	15	18	21	24	27	30	33	36
4	4	8	12	16	20	24	28	32	36	40	44	48
5	5	10	15	20	25	30	35	40	45	50	55	60
6	6	12	18	24	30	36	42	48	54	60	66	72
7	7	14	21	28	35	42	49	56	63	70	77	84
8	8	16	24	32	40	48	56	64	72	80	88	96
9	9	18	27	36	45	54	63	72	81	90	99	108
10	10	20	30	40	50	60	70	80	90	100	110	120
11	11	22	33	44	55	66	77	88	99	110	121	132
12	12	24	36	48	60	72	84	96	108	120	132	144

come up and find a duplicate number and decide if it is a result of a rotated array or if it is made from a different array. Introduce the term **commutative property** to describe these arrays and multiplication problems.

Ultimately, you want students to realize that what they have just created is a multiplication chart, something that is probably hanging somewhere in their classroom. At this point, this chart will have so much more meaning than a mere magic chart of numbers.

→ When exploring patterns on the multiplication chart, children will notice that there are mirroring numbers on the top and bottom of the chart, but will see that there are some numbers without a pair. These numbers are located down the center of the chart. Ask students why they think that they there is no pair. Ask students to build those arrays. They will soon realize that if they turn them, like the other arrays, the numbers are the same. This is a perfect time to introduce **square numbers.**

Resources for multiplication practice

→ Timez Attack: http://www.bigbrainz.com/ Claimed to be THE most successful way to master multiplication tables! They also have division, addition and subtraction. My children loved this and it really does work. There are free and pay versions. The free version provides the full range of mathematics and is all you need The pay version gives you access to more worlds.

→ For math timed tests and other computation support: http://www.hoodamath.com

→ Schoolhouse Rock Multiplication dvd/cd, also found on Youtube.

→ *Memorize in Minutes: The Times Tables* www.multiplication.com
This book targets the right brained learner and visual thinker by associating each multiplication fact to a silly picture and story.

Emphasizing the appropriate language of Multiplication

Show students a multiplication problem on the board, such as 3 x 4, and ask them what this means and to model using some type of counters that are available.

Children may very well show the problems as follows:

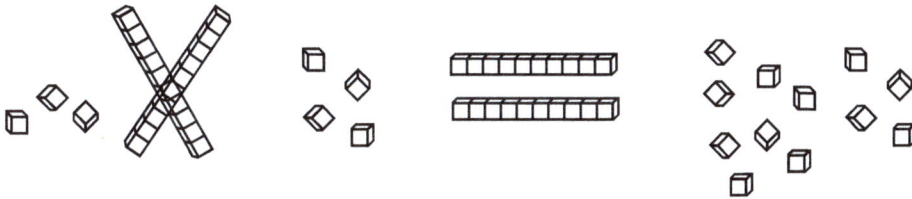

If they do, they are simply showing the symbolic equation in the form of blocks. This, or simply showing the answer, would tell you that they do not know how to use the blocks to show the problem.

Note: If you know that this is their initial experience with multiplication, such as the end of second or beginning of third grade, then you can skip this initial assessment and start with the following problem.

Now present the problem in context, using a problem such as:

> **"There are three friends and each one had 4 cookies each. How many cookies do they have altogether?"**

This time when students model, they will act out the problem, either literally, or with cubes. When describing the model, ask how they can represent this in a number sentence. Reinforce the language that there are three groups of 4. Students may either write 4 + 4 + 4 or 3 x 4, depending on their prior experience. More than likely in a group you will get a little of each. Help children make connections between the two. When reading 3 x 4, let them know that there are different ways to say it, but for purposes of understanding, you are going to say "groups of" for the "x" symbol.

Moving from groups to arrays to the area model
Build, draw, connect to numbers

Developmentally, children work in groups, or piles, before dealing with arrays. Sometimes they are not ready to move to arrays when the curriculum suggests. To help the transition, some students need more explicit connections between the two models. I find that the way to do this is to first pose a problem, such as the cookie problem on the previous page, and then have them arrange their cookies into straight lines and put them next to each other.

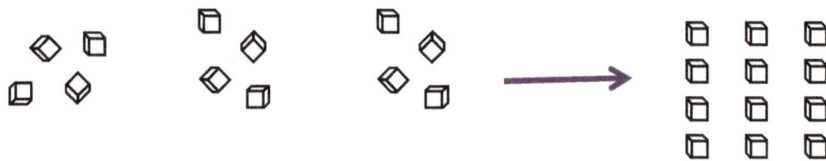

Ask the student (in the array) where each person's cookies are. This will help them to understand the organization of the array because it is not uncommon for them to loose the numbers in this way of organizing. Having them build groups then organize in rows and columns helps them to keep tabs on where the groups are.

Initially, students should be working with arrays where the objects are not touching each other, so they can still find their groups. When it is apparent that they are comfortable with this model, then they can model by using square tiles or cubes and they can shove them together to create an area model with no gaps or spaces.

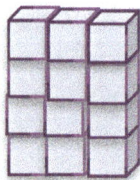

When the numbers are preserved in the concrete model, give the child a piece of cm or one inch square paper to outline what was built and label the numbers. Gradually transition from building and drawing to just using pictures to identify the multiplication sentence that matches.

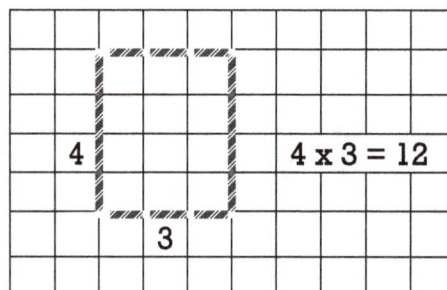

$4 \times 3 = 12$

Breaking apart: single digit problems
Build, draw, and connect to numbers

Materials: base ten blocks, grid paper, markers

The **distributive property** is basically pulling apart a multiplication problem, solving the parts and putting it back together. We want children to play with this idea with single digit numbers before applying it to larger ones. Using this concept can help those students who are having trouble memorizing their basic facts. Students that are more large-picture thinkers (right brain lead), tend to have descent reasoning skills, even if they are unable to memorize discrete facts.

Have students use square color tiles to build 6 x 8 and have them draw it on a grid paper.

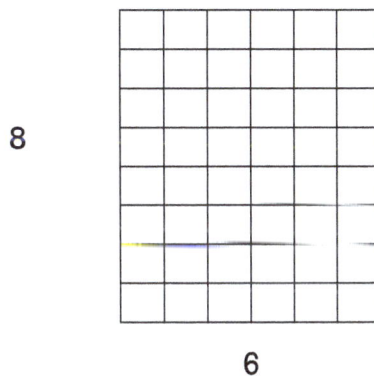

8

6

Ask them to take their tiles and split them into to two parts (they do not have to be equal). Initially, they will choose to split the 8 into 4 and 4, or into 3 and 3. Have them draw a line on their area model where they split their array and write the number sentences to match each section (i.e., 4 x 6 and 4 x 6, if they split the 8 into 4 and 4). Most likely students will split into two equal parts, so have them put it back together and then split again in a way that no one did it. Have them write their number sentences like the one below, and when ready they can use an open area model to record.

(4 x 6) + (4 x 6) =
  24 + 24 = 48

	4	4
6	4 x 6	4 x 6

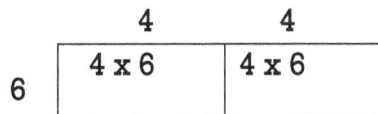

I have found that introducing the idea of breaking apart using the area model can be confusing for some and a better model is through using a counter to represent a group:

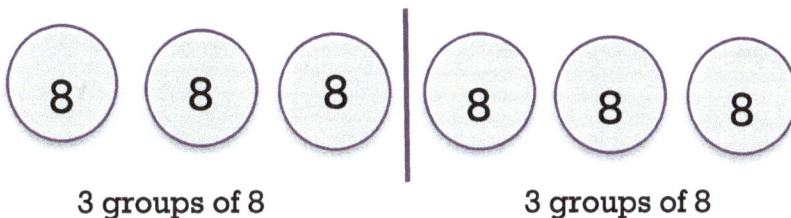

(8) (8) (8) | (8) (8) (8)

3 groups of 8   3 groups of 8

103

Multiplying by multiples of ten
And other patterns explored with number strings.

Once students are familiar with their basic facts, or as they are learning them, they can have experiences in multiplying multiples of ten. Allow students to initially use a calculator while exploring number strings so that they try to see what patterns they notice.

$$4 \times 3 =$$
$$4 \times 30 =$$
$$40 \times 30 =$$
$$400 \times 30 =$$
$$400 \times 300 =$$

Help students use base ten blocks or pictures to explore what is happening to the numbers. We want them to generalize the rule that the number of zeros in the factors will determine how many zeros will be in the product. Note that some factors, such as 5×6 yield products with zeros in them, so that will be an additional zero.

Some other multiplication principles to explore are:

→ What happens when one factor doubles and the other remains the same

6×3
12×3
24×3 **Student generated generalization:**
When one factor doubles and the other remains the same, the product doubles (can be easily shown with an array)

→ What happens when both factors double

6×3
12×6
24×12 **Student generated generalization:**
When both factors double, the product doubles (can be shown with an array).

→ What happens when one factor doubles and the other is cut in half

24×3
12×6
6×12 **Student generated generalization:**
When one factor doubles and the other is cut in half, the product remains constant.

Estimating

Using estimating is a way to expose students to larger numbers, without actually having to get exact answers. It also allows students to think about the reasonableness of an answer when solving for the exact answer. Estimating is a right-brained skill and may be hard for some students to step back and look at the bigger picture. There are two basic estimating strategies that we would like students to use, but we want them to choose an estimating strategy depending on the numbers presented. The act of deciding based on the numbers is not easy and takes a lot of practice. Therefore, teachers need to provide a lot of opportunities for this decision making process.

26 x 8

→Estimating by rounding

30 x 8 = 240. Many children will round the 8 to 10. Explore what happens the product when both factor are rounded. Since the point of rounding is to get to a point where they can use basic facts, it is unnecessary to round a single digit number.

→Estimating by using compatible numbers

25 x 8 = 200. Using compatible, or "friendly numbers" can yield more exact estimates, but requires more number sense to decide what numbers to choose. In this case, 25 would be a compatible number because most students know money and 8 quarters is $2.

Using compatible numbers to multiply

Another reason we want students to understand the idea of compatible numbers is because it is also a good way to getting the exact answer. Lets take the problem from the previous page as an example: 26 x 8. If students use compatible numbers and change the problem to 25 x 8, then they have to notice whether they took some away or added some extra groups. It is hard for them to visualize this and may think that they have to add 25, so they might need to draw it out to see it more clearly:

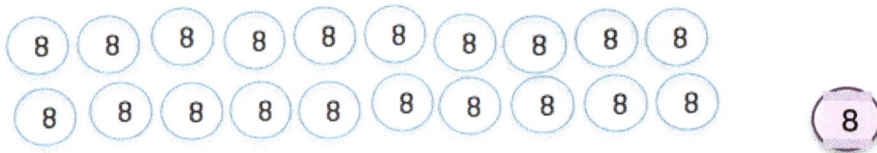

$$(25 \times 8) + (1 \times 8) =$$
$$200 \qquad 8 \quad = 208$$

In this picture, students can easily see that there is an 8 that needs to be added rather than a 25.

If the problem were 24 x 8 instead, the student may still want to use 25 x 8, but will see that they need to take away an 8 at the end since they had to add one to make the problem easier to solve.

$$(25 \times 8) - (1 \times 8) =$$
$$200 \quad - \quad 8 \quad = \quad 192$$

Materials: base ten blocks, grid paper, markers

Breaking apart: multi- digit problems
Build, draw, and connect to numbers

I usually start this experience by posing the following problem:

> I want to tile my patio. I measured it to see how many 1 foot tiles I need to buy from the local hardware store. My measurements are 6 by 10 feet. How many tiles do I need to buy?

Most all students will know that the answer is 60, but I make them build it with the base ten blocks. They will all pretty much get 6 ten sticks and say that the answer is 60. On the board I will draw what they built and label the area model.

10

6

Depending how much building experience students have had up to this point, there is a huge likelihood that they are only showing the answer and were not thinking at all about the dimensions.

I then pose the second problem:

> I went to measure the patio again to make sure and I found out that the exact measurements were 6 x 12 feet. How many tiles do I actually need to buy.

Again, depending on their experience, some students will know the answer (or solve with numbers on paper/in the air) and show the answer by presenting 7 tens and 2 ones. When they do so, I always ask if this looks like my patio and I refer back to the previous problem and the picture I drew. They realize that it doesn't. I also ask, in their model, where is the 6 and where is the 12. This will be enough to get them back to actually thinking about the dimensions and matching it to a model.

Once students are appropriately modeling the problem, I will have them record their model on grid paper and label it.

When I ask the student how many blocks they have altogether, they will instinctively say, "I know that this is 60," referring to the tens, "and this is 12," referring to the ones, "so that is 72 altogether.

On a sketch on the board I will draw their model and record the number sentences that go with the parts that they tell me.

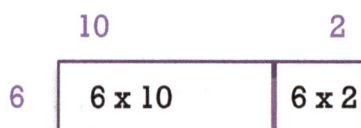

	10	2
6	6 x 10	6 x 2

$$(6 \times 10) + (6 \times 2) =$$
$$60 \quad + \quad 12 \quad =$$

Students need several opportunities to build, draw and record before asking them to start mentally multiplying 2 x 1 digit numbers, but afterwards, they are able to pull the tens and ones apart and multiply them to the second factor.

2 x 2 digit and beyond

Obviously the most common mistake students make when trying to multiply factors that are both 2 digit numbers is that they multiply the two ones and the two tens. They are applying addition rules to multiplication. This is why they need solid experiences and concepts around what multiplication is so that they can use either concrete models or pictures to make sense of what they have to do with the numbers.

You can take the above tiling scenario and now say that the dimensions are 16 x 12. When initially building this, you will notice students either using all ones or lots of tens. I let them go through the pain of being so tedious and then at the end, if they haven't already figured out that they can use a hundred flat for some of the tens, or they are using only unit cubes, I will ask if there is a way to use less pieces when making their model. This usually gets them to realize that they can trade for larger pieces.

Just like with single digit and 2 x 1 digit, it is important that the students build, draw and label, and record the partial products using number sentences. When they do, they will see that if they only multiply the ones and the tens, they are in fact missing two sections of the model.

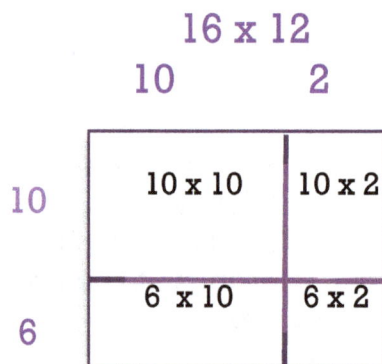

16 x 12

	10	2
10	10 x 10	10 x 2
6	6 x 10	6 x 2

$$(10 + 6) \times (10 + 2)$$

$$
\begin{array}{r}
10 + 6 \\
\times\ 10 + 2 \\
\hline
(10 \times 10) = 100 \\
(10 \times 6) = 60 \\
(10 \times 2) = 20 \\
(6 \times 2) = 12 \\
\hline
192
\end{array}
$$

Especially if students have already learned the standard algorithm, they need to understand that it is a short cut for this process of partial products. They first need to understand the long cut before they can use the short cut efficiently and with understanding.

Connecting to the standard algorithm

I am very careful before I introduce the standard algorithm to my students, especially if they are struggling in school. I clearly remember one year when I pre-maturely introduced the standard algorithm to my 4th graders. We had spent a lot of time building sense of multiplication and connected it to numbers. Students were usually drawing open area diagrams to help them solve their problems (actually indicating that they still relied on a picture for problem solving). The chapter test was going to be the next day and for some reason I felt inclined to show students the standard algorithm, when in fact I probably only had about 4 students who were truly ready. Well, needless to say, it confused them and since they somehow had the idea that it was the "right" way, this is the way I saw most of them attempt to use on their test to solve the problems. It wasn't pretty.

$$
\begin{array}{r}
16 \\
\times\ 12 \\
\hline
32 \\
+120 \\
\hline
152
\end{array}
$$

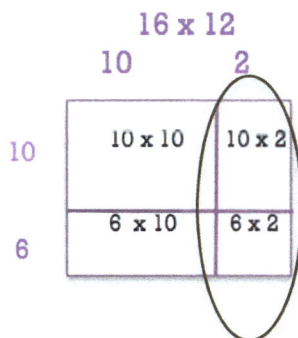

16 x 12

	10	2
10	10 x 10	10 x 2
6	6 x 10	6 x 2

When getting the children to make connections between the standard algorithm and the model, after solving, I have them find the numbers in the model. In this case the 32 is located in the right column and the 120 is on the left.

In addition, when multiplying 6 x 2, the product is 12. The next step is multiplying 10 and 2 then adding the additional 10 from the 12. In the partial products recording on the previous page, the 12 and 20 are added together, the 2 tens and 1 ten are added, similar to that of the standard algorithm, but the 1 ten is in a different location.

Division as inverse of multiplication

When posing story problems involving sharing amounts that are part of the basic facts (such as sharing 15 cookies with 3 children), many students will solve through multiplication: 3 x 5 = 15. This is not much different as when children are learning to subtract. If they do this, they recognize that sharing is opposite of equal groups. Students who do not know their multiplication tables, however, often find it hard to recognize this correlation and think that division facts are a whole new set of problems to memorize. This is why trying to get students to learn their facts early is so beneficial, so when it is time to divide, they have a resource for solving the problem.

Many students are able to use the multiplication chart to multiply, but they do not have an idea how to use it for division. If the problem is 56 ÷ 7, teach them to go down the 7's column and find the 56 and find out what row they are on. Even if it is a problem involving a remainder, such as 58 ÷ 7, they can go down the 7's column until they get to a number that is close 58 without going over, then figure out how many extra there are.

Modeling division with appropriate language

As with multiplication, we need to be using clear language that supports understanding of what is happening with division. Using phrases such as "shared with," and "split into," help emphasize that the total amount is being split into groups.

If your students have had prior experience with division, you may want to ask them to show you $6 \div 3$ using counters of some sort. If they are unable to model, or use ten sticks to represent the equal sign, tell them that the ten sticks represent and amount and cannot be in your model.

Let's consider the following problem:

> 24 marbles are shared equally with 4 boys.
> How many marbles does each boy get?

If this is the child's first experience with division, they may try to make 4 groups of 24, if thinking of multiplication. This recently happened in a 4th grade class after a hefty multiplication unit. They also may try to break the total into equal groups, but may not remember that it needs to be 4 groups. Instead they might be happy when they find that they can equally split the 24 into three groups. When this happens, I find it very helpful to provide students with index cards or paper plates so they can use them to represent the groups.

When describing their models, using language such as "shared with" rather than "divided by" helps children connect the act of sharing with dividing when writing the symbolic notation $24 \div 4 = 6$.

Partitive vs. quotative division models

There are two scenarios, or models, for division. Take the previous problem. To exactly model it, students would have to deal out cubes (or marbles) to each group. They may do it one at a time, two at a time, or 4 at a time, but they would still be dealing out since the total number of groups is known and the number in each group is the unknown. This problem type is called ***partitve*** division.

Take this problem in consideration:

> There are 24 marbles, and each boy receives 4 marbles.
> How many boys are able to get marbles?

In order to solve this one, students will put 4 into each group until they run out of marbles. This model is called ***measurement*** or ***quotative*** division. It is important that students have the opportunity to work with each type. In this model, division is acting as repeated subtraction.

As we work with larger numbers, we will actually use the partitive model to help derive an algorithm.

Exploring division patterns with number strings

Number Strings are a perfect way to explore patterns in the different operations and division is no exception. Some division principles to explore are:

→ What happens when the dividend is doubled:

$$6 \div 3$$
$$12 \div 3$$
$$24 \div 3$$

Student generated generalization:
When the dividend doubles, the quotient also doubles. If I have twice as much to split, I will have twice as much in each group.

→ What happens when both the dividend and divisor doubles (or is multiplied by 10, etc.)

$$6 \div 3$$
$$12 \div 6$$
$$24 \div 12$$

Student generated generalization:
When the dividend and divisor doubles, the quotient remains the same.

This big idea will be useful when getting into dividing with decimals and the need to "clear" the decimal, which is just multiplying both the dividend and divisor by either ten, hundred, etc., in order for the decimal to be removed.

→ What happens when the divisor doubles and the dividend remains the same?

$$24 \div 3$$
$$24 \div 6$$
$$24 \div 12$$

Student generated generalization:
When the divisor doubles and the dividend remains the same, the quotient is cut in half.

Developing an algorithm
Build, draw, and connect to numbers

As we did with the simpler division problems, allow students to use index cards or paper plates if they need support to remember their groups. By now, they may have weaned themselves from this scaffold.

Continue to give the division problems in context. As you progress, the number size will become larger and more challenging.

→ 2 digit dividends and one-digit divisors no remainders
→ 2 digit dividends and one-digit divisors with remainders
→ 3-4 digit dividends and one-digit divisors with remainders
→ multi-digit dividends and two-digit divisors with remainders
Note: when bridging to two-digit divisors, use estimation first as a strategy so they are dividing multiples of tens.

As students model how to solve their problems, they need a few days with just the blocks, then they need to be explicitly taught how to draw their pictures. They should NOT draw what is in each group after they have all been partitioned. Rather, they should draw as they go along. This will help them make sense later of their numbers.

Build and draw

$7\overline{)428}$

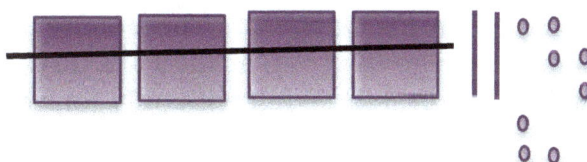

1 – after building 428, exchange the 4 hundreds for tens.

2 – partition a total of 6 tens into each group and 1 one.

3 – There is one left over

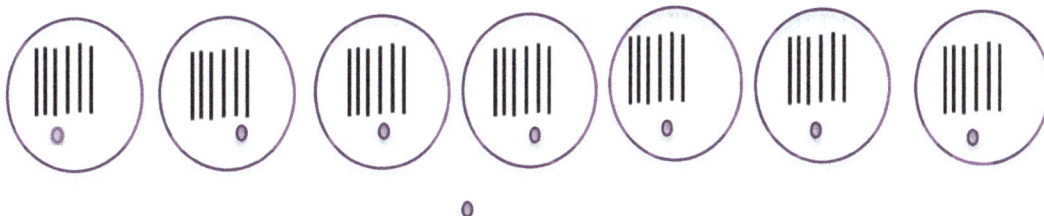

Connect to an algorithm

When ready to explore how to record this with numbers, just as when drawing, students should be recording as they go along, not all at the end. They should build a step, draw a step, and record a step with numbers.

```
          61 R 1
      ┌─────────
    7 │ 428
        140    20
        288    20
        140
        148    20
        140
          8     1
          7
          1
```

In the previous model, the student partitioned the blocks into the 7 groups. In this case, let's say he partitioned 2 tens into each group. Generally students either choose one or two tens to start. If that were so, then he would have put 20 in each group, getting rid of 140 from the pile. We will record the 20 and 140 like this:

This student finds out that there are 288 left in his pile and chooses to partition another 20, leaving him with 148. He could easily see that there was enough to put in an additional 20 in each group, leaving him with 8. He could see that he can only put one into each group and he would have one left over.

When he is all done, he sees that he has 61 in each group with one left over.

This strategy is called *Partial Quotients*. The beauty of using this strategy is that children can chunk as small or large of group as they are developmentally ready to. They also could partition the ones first to be left with only tens or hundreds.

The ultimate question is, "how many 7 can make 428?" When moving beyond using the cubes, students may want to ask themselves, how many 7's can make 400 or 420? The goal is to see if they can recognize a basic fact or to get close to it. They should ask themselves, "could there be at least 100? 10?" If the answer is yes, then ask, "could there be 2 tens, 3 tens?" This helps students start thinking in larger chunks. Of course there will always be those who go right to their times tables, but the students that you are most likely to be working with if you are giving this assessment is a struggling learner, who will not be thinking much of his facts.

Connect to the standard algorithm

The rule of thumb I use when teaching is I tell my students they cannot use a strategy they do not understand. This generally means the standard algorithm. Most children end up in tutoring in 4th and 5th grade as a result of not understanding long division. Students are dealing with individual digits and bringing down numbers, often even drawing arrows…they have no idea what is going on in that algorithm. I personally only use the partial quotients algorithm when I divide because it is efficient and I don't have to worry about trial and error. However, I now that our students are exposed to the standard algorithm, either with other teachers, siblings, parents, online computer aides, etc. It then becomes important in helping them make connections between what they are doing and the standard algorithm.

$$
\begin{array}{r|l}
& \underline{61}\,R\,1 \\
7\,)\,\overline{428} & \\
\underline{140} & 20 \\
288 & 20 \\
\underline{140} & \\
148 & 20 \\
\underline{140} & \\
8 & 1 \\
\underline{7} & \\
1 &
\end{array}
$$

Less efficient way
With partial quotients

$$
\begin{array}{r|l}
& 61\,R\,1 \\
7\,)\,\overline{428} & \\
\underline{420} & 60 \\
8 & 1 \\
\underline{7} & \\
1 & 1
\end{array}
$$

Most efficient way –
with partial quotients

$$
\begin{array}{r|l}
& 61\ \ R\,1 \\
7\,)\,\overline{428} & \\
\underline{42} & \\
8 & \\
\underline{7} &
\end{array}
$$

Standard Algorithm
(short cut)

Essentially, doing the standard algorithm is what a student would do in the partial quotients method if using the most efficient strategy. The difference is that the value of the number is preserved in the partial quotients way and students are not thinking of isolated digits without meaning.

Order of Operations

Order of operations is a convention that is not based on conceptual understanding, so rules have to be taught to students. Simple calculators do not calculate based on the rules of order of operations, but scientific calculators do. Therefore, when teaching these rules, it is important that students experience solving problems with both types of calculators so they can see the difference.

Depending on the level of order of operations that they are working, they need to understand the hierarchy of what needs to be solved first. For example, they need to understand that multiplication and division is more complex than addition and subtraction, and that exponents are more complex than multiplication and division. They also need to realize that what is grouped in the parenthesis need to be solved before they can be calculated with the rest of the number sentence. I think that deep discussions regarding this hierarchy are more effective than the common chants of PEMDAS.

A common misconception that comes up, especially when using pneumonic devices, is that students tend to think that multiplication comes before division and addition comes before subtraction. Therefore, they may be found solving all the multiplication from left to right, then go back to find the division, rather than treat multiplication and division as being of equal importance.

Parenthesis – target the grouped sections first
Exponents – repeated multiplication
Multiplication / division - repeated forms of addition and subtraction

Skip Counting Charts

2	4	6	8	10	12	14	16	18	20	22	24	26	28
3	6	9	12	15	18	21	24	27	30	33	36	39	
4	8	12	16	20	24	28	32	36	40	44	48	52	
6	12	18	24	30	36	42	48	54	60	66	72	78	
7	14	21	28	35	42	49	56	63	70	77	84	91	
8	16	24	32	40	48	56	64	72	80	88	96	104	
9	18	27	36	45	54	63	72	81	90	99	108	117	

I can skip count by: (check when you learn them)
2's__   3's__   4's__   5's__   6's__   7's__   8's__   9's__10's__

Multiplication Tables
Highlight the ones you know & study the rest.
Have someone else help test you. (Test your facts in random, not sequential, order)

X	1	2	3	4	5	6	7	8	9	10	11	12
1	1	2	3	4	5	6	7	8	9	10	11	12
2	2	4	6	8	10	12	14	16	18	20	22	24
3	3	6	9	12	15	18	21	24	27	30	33	36
4	4	8	12	16	20	24	28	32	36	40	44	48
5	5	10	15	20	25	30	35	40	45	50	55	60
6	6	12	18	24	30	36	42	48	54	60	66	72
7	7	14	21	28	35	42	49	56	63	70	77	84
8	8	16	24	32	40	48	56	64	72	80	88	96
9	9	18	27	36	45	54	63	72	81	90	99	108
10	10	20	30	40	50	60	70	80	90	100	110	120
11	11	22	33	44	55	66	77	88	99	110	121	132
12	12	24	36	48	60	72	84	96	108	120	132	144

Timed Test Score Card

http://www.hoodamath.com/mobile/games/mathtimedtests.html

Date	Circle one	% correct	Time
	+ - x ÷		
	+ - x ÷		
	+ - x ÷		
	+ - x ÷		
	+ - x ÷		
	+ - x ÷		
	+ - x ÷		
	+ - x ÷		
	+ - x ÷		
	+ - x ÷		
	+ - x ÷		
	+ - x ÷		
	+ - x ÷		
	+ - x ÷		
	+ - x ÷		
	+ - x ÷		
	+ - x ÷		
	+ - x ÷		
	+ - x ÷		
	+ - x ÷		
	+ - x ÷		
	+ - x ÷		
	+ - x ÷		
	+ - x ÷		
	+ - x ÷		
	+ - x ÷		
	+ - x ÷		
	+ - x ÷		
	+ - x ÷		
	+ - x ÷		
	+ - x ÷		
	+ - x ÷		
	+ - x ÷		
	+ - x ÷		

⁺Fractions

Assessment

Present each pre-printed question card to the child. Only present as many of the questions as the child shows understanding except for the first question, *identifying a half*. I find that they may still be able to answer the next question or two, even if they do not have full understanding of ½. Therefore, you can still proceed if the student gives some evidence of understanding of ½.

To save paper, laminate or present student task cards in plastic page protectors so students can use a dry erase marker to record their work or answers. Record student work on the worksheet. See worksheet for additional instructions for specific questions were applicable.

Scoring
Mark the box in front of the question if they were accurate and record the types of responses made. Each mark is worth one point. You may choose to give ½ credit in certain circumstances. Mark if it appears that students have conceptual understanding versus just solving through a procedure. If they go straight to a numerical strategy, ask them if they are able to show their strategy using a picture, or to explain their thinking in a way that would indicate that they have conceptual understanding.

Note: There are many levels of fractions, more than which are assessed (see pages 134 & 136). This assessment will give you an idea as to how students deal with comparing and operating some, but not all, levels of fraction.

1. Circle each figure that shows ½

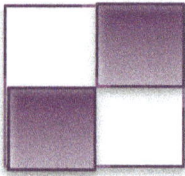 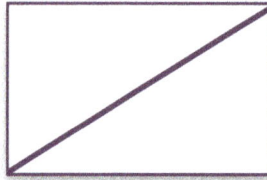 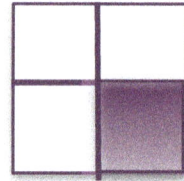

 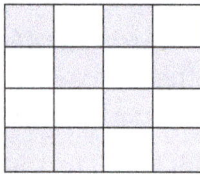 

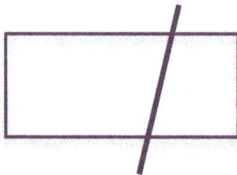 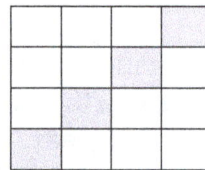

2. This desert plate has brownies and peanut butter cookies. Kim said that the brownies make up 2/3 of the plate. Sean said that they make up 2/5. Who is correct and how do you know?

3. How many fractions can you find in the diagram? Label as many as you can.

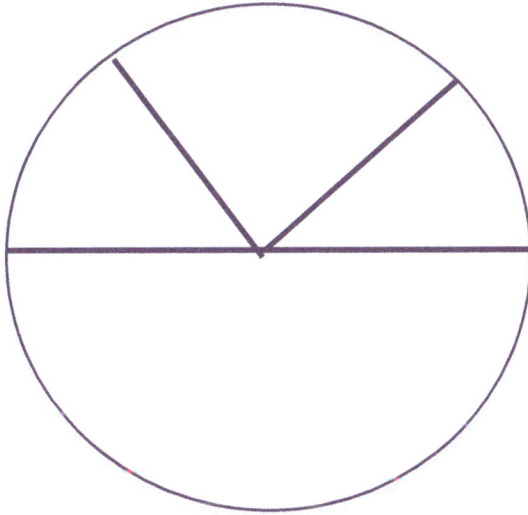

4.

5. The library is ¾ of a mile from my house. The grocery store is 6/8 of a mile from my house. Which one is farther from where I live? Show your thinking.

6. Which fraction is larger?

    a.   1/2   2/5

    b.   3/8   4/5

7. Four brothers went home for an after school snack. Their mother prepared 3 sandwiches. Show how these boys are able to split up the sandwiches so each can get the same amount. How much of a sandwich can each child get?

8. When making lemonade for their lemonade stand, Joshua and Andrew wanted to make 10 liters of lemonade. For each liter, they used ¼ juice to ¾ water. How much juice and water did they need for the 10 liters?

9. Susan and Kim want to make brownies for the bake sale. The brownies call for ¾ cup of cocoa. They want to double the recipe and are trying to decide how much cocoa they will need. Susan says they will need 6/8 cup of cocoa and Kim says they need 6/4. Who is correct and how do you know?

10. Will the sum be 1, less than 1 or greater than 1? How do you know? Solve to check.

$$1/2 + 2/3$$

11. Will the difference be ½ , less than ½ , or greater than ½ ? How do you know? Solve to check.

$$5/8 \ - \ 1/4$$

12. Margaret made cupcakes for the bake sale. She made 3 different kinds. Each kind she used 2/3 cup of frosting. How much frosting did she use altogether?

13. Will the product be larger or smaller than ½? How do you know? Solve to check.

1/2 x 2/3

---

14. Adrian lives 2 ½ miles from school. If every ¼ mile there is a stop sign, how many stops signs are there between his house and school?

15. Will the quotient be greater or less than 3? How do you know? Solve to check.

$$3 \div 1/2$$

Interventions for teaching fractions

The breadth of the fraction concepts tested spans most of the elementary school grades. Therefore, it is not within the scope of this manual to provide a comprehensive list of intervention strategies to teach all areas tested in this assessment. This assessment serves to be a guide as to where the students are developmentally and can help pinpoint misconceptions and solution strategies. There are, however, some key activities that I wanted to point out that can help children make some real sense of fractions. I have included these on the following pages. I am also including a list of additional resources, which are very supportive in deepening both the students and teachers' understanding of fractions.

Concept of ½

<u>Quilt Squares</u>

The purpose of this activity is for students to see that one-half is not just a whole split into two parts, but that there can be many parts and as long as there is an equal number of shaded and un-shaded, for example, then the shaded sections total 1/2.

1 - Using the quilt squares provided on the next page, have students find as many ways to color in 1/2. They should use a different square for each way.

2 – Trim squares. Choose the amount of squares necessary to cover a piece of black construction paper.

3 – Arrange the squares to make a pleasing design. Glue squares to make your quilt.

<u>Extensions</u>

Make designs for 1/4.

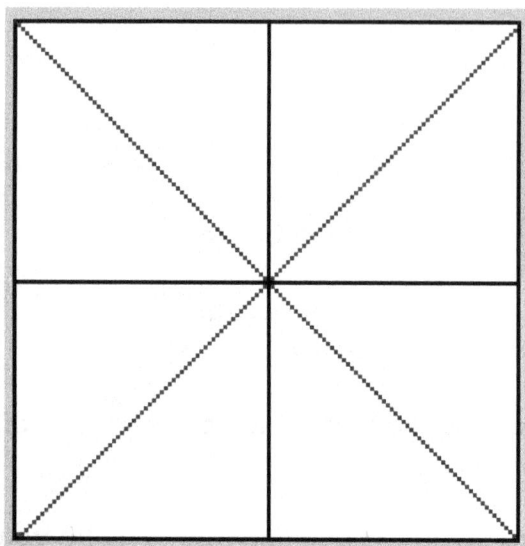
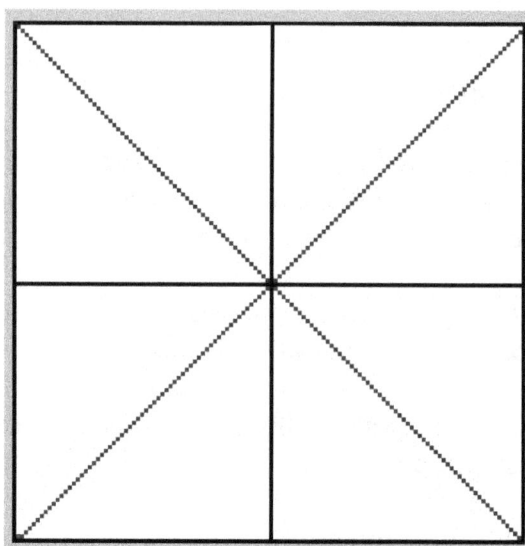
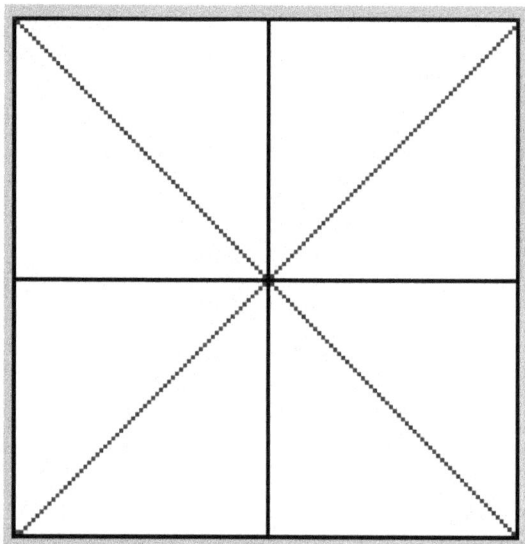
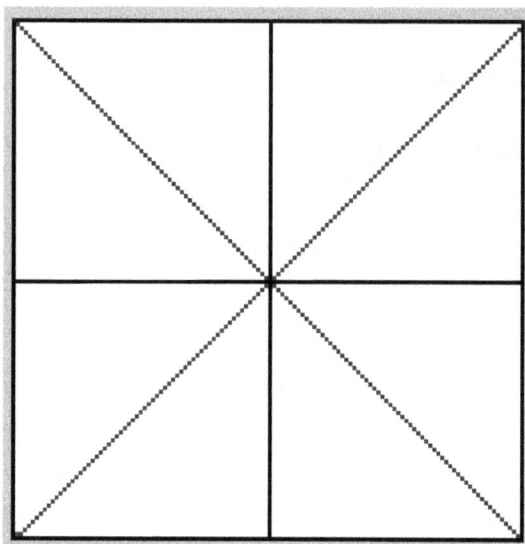
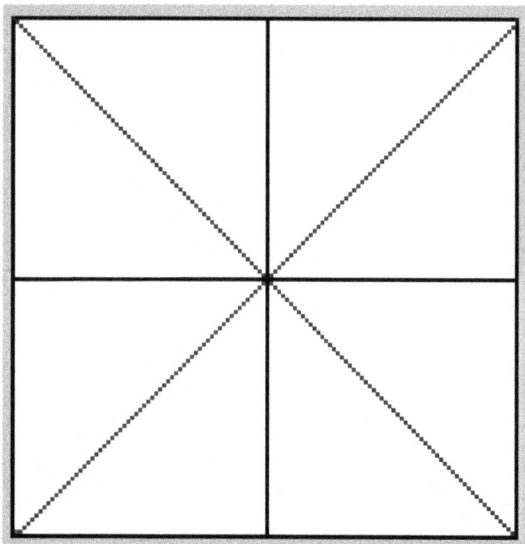
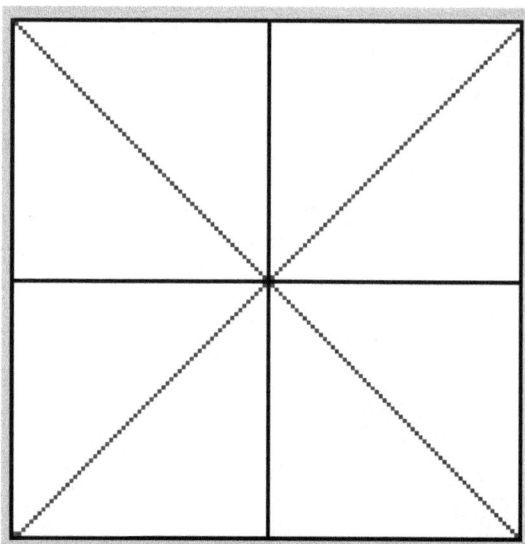

132

Equivalent Fractions

When engaged in activities to explore equivalent fractions with different manipulatives, such as fraction circles, fraction strips, Cuisenaire™ Rods, double sided counters, etc., it can be helpful to create charts where students come up with as many different ways to make/write a fraction and write it on the chart in the form of a collage. Those charts can be hung in the room for reference. I have found that the act of creating the charts lead students to develop generalizations such as, 'the denominator is double the denominator', because they have to think past the basic equivalent fractions that they already know. These generalizations are usually subconscious, so teachers need to ask questions such as, "How did you know that 100/200 would equal ½?" These generalizations, or rules, should be written on a sentence strip and stapled to the bottom of the chart.

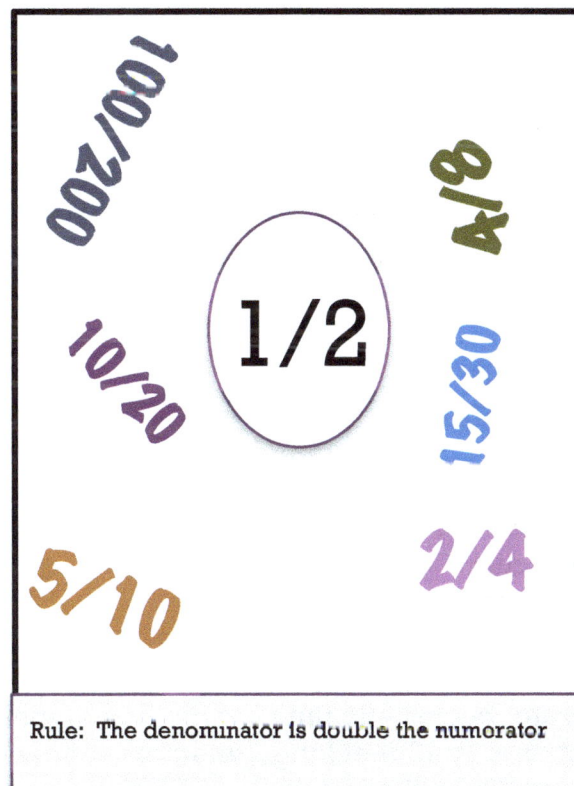

100/200   8/4

1/2

10/20   15/30

5/10   2/4

Rule: The denominator is double the numerator

Fractions on a number line

Knowing where fractions lie on a number line is more abstract concept than part-whole or part-set fractions. We may often underestimate the leap students have to take when learning how to order numbers on a number line. One reason is that even though they may have experience using fraction strips, the strips are marked with the fraction in the middle of the strip. However, on a number line, the marks are made at the end of the segments.

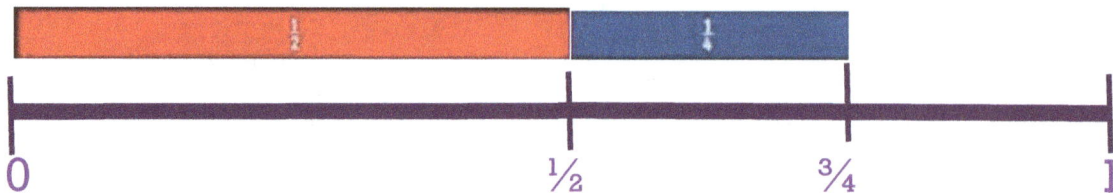

Students need to understand that ½ is not simply a point on the number line, but rather the amount of space from zero up to that point. Another misconception is that for the ¾ mark they will say ¼ because the amount of space from ½ to ¾ is ¼ long.

Building the Number line

Identify something long in the classroom, hallway, or on the playground and have students decide how to figure out where ½ is exactly on that object. This could be a white board, marked off section of the hall, or playground equipment. In my room I have a balance beam, which we used.

At first children want to eyeball it, but make them prove it. They try by finding objects around the class to reiterate, but usually they cannot find anything that will end up exactly even with the beam. It is exactly 12 feet long, yet interestingly enough no one starts off by suggesting using rulers. Once they use rulers, I make sure that I provide less than they are able to spread along the entire length so they have to come up with strategies for measuring. I notice that often the notion of making sure the rulers touch end to end is not there and have to discuss what happens with the accumulated gaps.

Once they accurately determine where the ½ is, then I have them find the 4ths, thirds, eighths, etc. I mark them all with post-it™ notes. This lesson takes several days.

Open Number Line

For this mathematical routine draw an open, or blank, number line on the board and label the ends with numbers such as 0 and 1 or 0 and 2. It depends on the level of numbers on which the work is focused. Give several students a sticky note with a number written on it, and let them place their number on the line where they feel it should go. When all the numbers have been placed, have everyone sit back, examine the placements, and discuss whether they agree or not. Or, you can have the class discuss after each placement. If the number line is completely open, with no marking for an end or beginning, then have one student at a time come up to the board and give her a number to place where she thinks it should go. Ask a second student to come up and, in relation to where the first number was placed, place a second number. In this scenario, students do not know all the numbers up front. After a few turns, someone needs to move numbers over, either to make room for new numbers or to redistribute the spacing for the number line to make sense. All students must explain why they put the number where they did. This activity is good for reasoning skills.

1/2   1/4   3/4   2/5   2

Computation with fractions

Like computation with whole numbers, students need to explore computation with fractions first using manipulatives then move toward generalizing rules.

<u>Addition and Subtraction</u>

Take this situation into consideration:

> In the refrigerator there is a ½ of a pepperoni pizza and ¼ of a sausage pizza. How much pizza is altogether?

When students use manipulatives to match, they might use fraction circles when putting together they may visually recognize the piece to be 3/4. But to prove it, they would have to convert the 1/2 to 2/4 and add it together with 1/4.

After several problems, and maybe even several days, a strategy students would generalize would to convert the fractions so that they are all the same color, or same name (i.e., fourths).

When providing addition and subtraction problems regarding fractions, consider the following problem types, in order from most basic to most concrete (Bahr & de Garcia 2010, p. 326-327).

Level	Type
A	Addition of like denominators $1/5 + 2/5 = 3/5$
B	Subtraction of like denominators $3/5 - 1/5 = 3/5$
C	Addition of mixed numbers with like denominators, with no regrouping $1\ 3/5 + 2\ 1/5 = 3\ 4/5$
D	Subtraction of mixed numbers with like denominators, with no regrouping $3\ 4/5 - 2\ 1/5 = 1\ 3/5$
E	Addition of mixed numbers with like denominators, with regrouping $1\ 4/5 + 2\ 3/5$
F	Subtraction of mixed numbers with like denominators, with regrouping $4\ 1/5 - 2\ 3/5 = 1\ 3/5$
G	Addition or subtraction of fractions with unlike denominators, for which one denominator is a common multiple of the other ½ $+ 1/6 = 3/6 + 1/6 = 4/6$ or $2/3$
H	Addition or subtraction of fractions with unlike denominators. The least common multiple of the denominators is less than the result of multiplying the denominators. ¼ $+ 1/6 = 3/12 + 2/12 = 5/12$
I	Addition or subtraction of fractions with unlike denominators. The least common multiple of the denominators is equal to the result of multiplying the denominators. ½ $+ 1/3 = 3/6 + 2/6 = 5/6$
J	Addition or subtraction of mixed fractions with unlike denominators with or without regrouping

Multiplication and division

Using the appropriate language, rather than "times" or "goes into" will help support understanding of what is happening when multiplying and dividing with fractions. It is also essential to go back to what is happening with whole numbers to make sense of what is happening when operating with fractions.

For example	2 x 3 means two groups of 3
So	2 x 1/3 means two groups of 1/3
and	½ x 1/3 means one-half group of 1/3

Another example	6 ÷ 3 means how many 3's in 6
So	6 ÷ 1/3 means how 1/3's make 6
and	¾ ÷ ½ means how many ½'s make ¾

Take these situations into consideration? How would you solve them?

A - I made 3 cakes. Each cake called for ¼ cup of milk. How much milk did I use altogether?

¼ + ¼ + ¼ = ¾   or   3 x ¼ = 3/4

B - I made some scarves, each scarf used 1/4 yard of material. I had 2 yards of fabric. How many scarves was I able to make?

2 ÷ ¼ = 8

Connecting multiplication and Division:

Take this example:

2/3 cup of water. I remove half of the amount. How much do I have left?

Students could solve this problem by saying 1/2 of 2/3, or 1/2 x 2/3. Or, they can think of it as 2/3 ÷ 2. If you notice, these two number sentences are the inverse of one another. It is through this understanding that students should develop the idea of invert and multiply. However, it needs to be noted that inverting and multiplying is not always necessary. Students with enough fraction sense can often divide mentally through reasoning.

When providing multiplication problems regarding fractions, consider the following problem types, in order from most basic to most concrete (Bahr & de Garcia 2010, p. 328-329).

Level	Type
A	Fraction with a numerator of one, times a whole number. The denominator is a factor of the whole number. ½ x 8 = 4
B	Fraction with a numerator of one, times a whole number. The denominator is not a factor of the whole number. ½ x 7 = 3 ½
C	Fraction with a numerator greater than one, times a whole number. The denominator is a factor of the whole number. ¾ x 24 = 18
D	Fraction with a numerator greater than one, times a whole number. The denominator is not a factor of the whole number. 2/3 x 3 = 1 ½
E	Whole number times a fraction. The denominator is a factor of the whole number. 8 x ½ = 4
F	Whole number times a fraction. The denominator not a factor of the whole number. 5 x ½ = 2 ½
G	Fraction times a fraction, with 1 in both numerators. 1/3 x ½ = 1/6
H	Fraction times a fraction, with 1 in only one numerator. ¼ x 2/5 = 2/20 or 1/10
I	Fraction times a fraction, with numerators other than one. ¾ x 2/5 = 6/20 or 3/10

When providing division problems regarding fractions, consider the following problem types, in order from most basic to most concrete (Bahr & de Garcia 2010, p. 332-333).

Level	Type
A	Whole number divided by a fraction 3 ÷ ½ = 6
B	Fraction divided by a fraction. The divisor is smaller, with a whole number as the result. ½ ÷ ¼ = 2
C	Fraction divided by a fraction. The divisor is smaller, with a mixed number as the result. ½ ÷ 1/3 = 1 ½
D	Fraction divided by a fraction. The divisor is larger, with the denominator of the divisor being a factor of the other denominator. 1/6 ÷ ½ = 1/3
E	Fraction divided by a whole number. ½ ÷ 3 = 1/6

Additional Resources for teaching fractions

The following are some valuable resources for learning about and teaching fractions:

Beyond Pizzas and Pies: 10 essential strategies for supporting fraction sense by McNamara & Shaughnessy.

Elementary Mathematics is Anything but Elementary by Bahr & de Garcia

Minilessons for Operations with fractions, decimals and percents, by Imm, Fosnot & Uittenbogaard.

Young Mathematicians at Work: Constructing Fractions, Decimals, & Percents Fosnot & Dolk

References & Resources

Bahr, D. and L. de Garcia. (2010). *Elementary mathematics is anything but elementary: Content and methods from a developmental perspective.* Belmont, CA: Wadsworth.

Chapin, S. H. (2003). *Classroom Discussions.* Sausalito, CA: Math Solutions Publications.

Fosnot, C. T., and M. Dolk. (2001). *Young Mathematicians at work: Constructing multiplication and division.* Portsmouth, NH: Heinemann.

Fosnot, C. T., and M. Dolk. (2001). *Young Mathematicians at work: Constructing number sense, addition and subtraction.* Portsmouth, NH: Heinemann.

Fosnot, C. T., and M. Dolk. (2001). *Young Mathematicians at work: Constructing fractions, decimals and percents.* Portsmouth, NH: Heinemann.

Hendrickson, S., S. C. Hilton, D. L. Bahr. (2009). Using the Comprehensive Mathematics Instruction (CMI) Framework to Analyze a Mathematics Teaching Episode. *Utah Mathematics Teacher. 2*(1).

McNamara, J., & M. Shaughnessy. (2010). *Beyond Pizzas & Pies: 10 Essential strategies for supporting fraction sense.* Sausalito, CA: Math Solutions Publications.

Parrish, S. (2010). *Number talks: Helping children build mental math and computation strategies.* Sausalito, CA: Math Solutions Publications.

Richardson, K. (1999). *Developing Number Concepts. Book 1: Counting, comparing, and pattern.* Parsippany, NJ: Dale Seymour.

Richardson, K. (1999). *Developing Number Concepts. Book 2: Addition and subtraction.* Parsippany, NJ: Dale Seymour.

Richardson, K. (1999). *Developing Number Concepts. Book 3: Place value, multiplication, and division.* Parsippany, NJ: Dale Seymour.

Sousa, D. A. (2008). *How the brain learns mathematics.* Thousand Oaks, CA: Corwin Press.

Treacy, K., & J. Cairnduff. (2009). *Revealing what students think: Diagnostic tasks for fractional numbers.* Ascot, Western Australia: STEPS Professional Development.

Willis, S., S. Devlin, L. Jacob, B. Powell, D. Tomazos, & K. Treacy. (2010). *First Steps in mathematics: Number.* Melbourne: Rigby.